3rd Pak-Turk International Conference on Emerging Technologies in the Field of Sciences and Engineering (ETSE 2020)

Held online due to COVID-19

IOP Conference Series: Materials Science and Engineering
Volume 899

Topi, Pakistan
9 - 10 June 2020

Editors:

Jameel-Un Nabi
Shamsher Sadiq
Zahid Halim

ISBN: 978-1-7138-2812-9
ISSN: 1757-8981

Printed from e-media with permission by:

Curran Associates, Inc.
57 Morehouse Lane
Red Hook, NY 12571

Some format issues inherent in the e-media version may also appear in this print version.

Institute of Physics
Dirac House, Temple Back
Bristol BS1 6BE UK

Phone: 44 1 17 929 7481
Fax: 44 1 17 920 0979

techtracking@iop.org

Additional copies of this publication are available from:

Curran Associates, Inc.
57 Morehouse Lane
Red Hook, NY 12571 USA
Phone: 845-758-0400
Fax: 845-758-2633
Email: curran@proceedings.com
Web: www.proceedings.com

TABLE OF CONTENTS

ASTROPHYSICAL S-FACTOR FOR RADIATIVE CAPTURE OF PROTON BY 13C AT LOW ENERGY 1
Abdul Kabir, B. F. Irgaziev, Jameel-Un Nabi

AN OVERVIEW OF METHODS FOR INVESTIGATION OF AEROELASTIC RESPONSE ON HIGH ASPECT RATIO FIXED-WINGED AIRCRAFT 6
Muhammad Khizer Ali Khan, Ali Javed, Nafees Mumtaz Qadri, Mohtasshim Mansoor, Farrukh Mazhar

IMPLEMENTATION OF DIFFERENTIAL TRANSFORM METHOD (DTM) FOR LARGE DEFORMATION ANALYSIS OF CANTILEVER BEAM 19
Hassaan Abbasi, Ali Javed

ON BOARD INTELLIGENCE FOR PUBLIC TRANSPORTATION IN DEVELOPING COUNTRIES USING IOT 28
A Rehman, T Mumtaz

NEUTRINO ENERGY LOSS RATES DUE TO 66-71NI IN STELLAR MATTER 35
Fakeha Farooq, Ramoona Shehzadi, Jameel-Un Nabi

SELF-CLEANING OF GLASS SURFACE TO MAXIMIZE THE PV CELL EFFICIENCY 40
Adnan Ayaz, Hussain Ahmad, Faraz Ahmad, Ahmad Khan, S. M. hasnain Tarmazi, Rizwan M. Gul, Saim saher

COST-EFFECTIVE, RELIABLE, AND PRECISE SURFACE MOUNT DEVICE (SMD) ON PCBS 55
B Iftikhar, M M Malik, S Hadi, O Wajid, M N Farooq, M M Rehman, A K Hassan

SOLAR THERMAL OPPORTUNITIES AND CHALLENGES IN PAKISTAN 60
Adnan Ayaz, Faraz Ahmad, Saim Saher, Muhammad Noman

DATA BASED SENSING OF SHALE OIL YIELD IN OIL SHALE RETORTING PROCESS 80
Hasan Qayyum Chohan, Iftikhar Ahmad

NOVEL SIMPLIFIED BLADE HUB LINE OPTIMIZATION OF HIGH-PRESSURE RATIO CENTRIFUGAL COMPRESSOR DESIGN USING NUMERICAL SIMULATIONS 89
A Khan, F Ahmad, S Ali, M R Siddiqi, M Z Ijaz

HOME SECURITY AND AUTOMATION BASED ON INTERNET OF THINGS: A COMPREHENSIVE REVIEW 100
Saman Fatima, Naila Aiman Aslam, Iqra Tariq, Nouman Ali

INVESTIGATION OF SENSORS & ACTUATORS BASED ON HANKEL SINGULAR VALUES 112
Sheharyar Malik

STRESS ANALYSIS OF COMPOSITE LEAF SPRING – COMPARATIVE APPROACH 119
Sheharyar Malik, Kamran S. Afaq

PREDICTING COMPRESSIVE STRENGTH OF CRM SAMPLES USING IMAGE PROCESSING AND ANN 127
M I Waris, J Mir, V Plevris, A Ahmad

EXPERIMENTAL STUDY OF RAILWAY BRIDGE SCOURING AND PROPOSAL OF REMEDIAL MEASURES-A CASE STUDY OF BRIDGE NUMBER 30, HARROW RIVER. 134
Muhammad Sohail Javed, Naeem Ejaz

SIGN LANGUAGE TRANSLATION IN URDU/HINDI THROUGH MICROSOFT KINECT 139
Sohail, M Y Khan, N A Siddiqui, Kiran

DEVELOPING NON- LINEAR RELATIONSHIP AMONG FACTORS AFFECTING THE RUTTING SUSCEPTIBILITY OF ASPHALT MIXTURES USING TWO PARAMETER WEIBULL DISTRIBUTION 146
Fatima Ashfaq, Imran Hafeez, Sabahat Hussan

INDIGENOUS DESIGN AND DEVELOPMENT OF SPLIT HOPKINSON PRESSURE BAR (SHPB) TEST SETUP FOR CHARACTERIZATION OF MATERIALS AT HIGH STRAIN RATES 165
M Bilal Nutkani, Muhammad Abid, Riffat Asim Pasha, Uzair Ahmed Dar

Author Index

IOP Conf. Series: Materials Science and Engineering **899** (2020) 011001 doi:10.1088/1757-899X/899/1/011001

3rd Pak-Turk International Conference on Emerging Technologies in the field of Sciences and Engineering 9-10th, June 2020

Conference Proceedings

9-10th June 2020 GIK
Institute, Topi, Pakistan

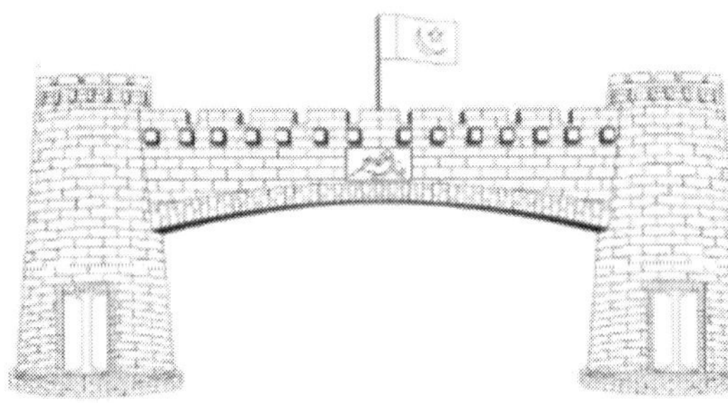

Ghulam Ishaq Khan Institute of Engineering Sciences & Technology, Topi

IOP Conf. Series: Materials Science and Engineering **899** (2020) 011001 doi:10.1088/1757-899X/899/1/011001

3rd Pak-Turk International Conference on Emerging Technologies in the field of Sciences and Engineering 9-10th, June 2020

Table of Contents

Preface
Aim and Scope of the Conference
Organizing Committees
Editorial Board
Program Schedule

IOP Conf. Series: Materials Science and Engineering **899** (2020) 011001 doi:10.1088/1757-899X/899/1/011001

3rd Pak-Turk International Conference on Emerging Technologies in the field of Sciences and Engineering 9-10th, June 2020

Preface

The Pak-Turk conference series is the technical event which focuses on the advances in new technologies in engineering sciences, nuclear astrophysics, renewable energy resources, optical engineering, computer science and electrical engineering. The purpose of this conference series is to provide a platform for researchers, academicians and practitioners to make them familiar with recent advancements in the various fields of engineering and sciences. The conference accepts wide range of papers to encourage young and experienced researches to present their work and the possibility of initiating mutual collaboration with international repute researchers and industry personals. The conference format comprises of multiple sessions and the selected research articles are presented based on substantial and novel research.

The series of event was initiated in 2018, by organizing "First International Pak-Turk workshop on Emerging Technologies in the Field of Sciences and Engineering" held on 9th May 2018 at Barber's Building on campus. Keeping in view the success of 1st and 2nd Pak-Turk international conference, this time, GIK Institute of Engineering Sciences and Technology organized the 3rd Pak-Turk Conference on Emerging Technologies in the Field of Sciences and Engineering on 9-10th June 2020, at GIK Institute. In view of the prevailing situation worldwide regarding Covid-19 and to comply with the instructions from the Government of Pakistan and to ensure the safety of participants the conference was conducted in virtual online mode of presentation. The conference comprised of Turkish, Pakistani and German speakers from around the globe. Various experts and scholars who work in diverse fields participated in the conference, expressed their views and spoke on different topics related to engineering education and scientific developments. Speakers in the two-day International Conference encouraged Pakistan to catch up the new emerging trends in the field of science and technology, as this is vital for the progress and economic prosperity. It was a fully synchronous online experience due to the COVID-19 outbreak; that had live Q&A, participants jointly listened to presentations, and attended other live events associated with the conference. The conference accepted only 15% of the submitted articles for publishing in the IOP proceedings. The online conference was attended by almost 200 plus individuals through the online video conference software. Furthermore, the conference was streamed live on Facebook making it visible and accessible to everyone, and hence was able to capture an audience of 3k plus around the globe.

The Conference proceedings and all the other scientific presentations related to the 3rd Pak-Turk ETSE-2020, must be attributed to the extraordinary efforts

IOP Conf. Series: Materials Science and Engineering **899** (2020) 011001 doi:10.1088/1757-899X/899/1/011001

made by the conference Scientific Publication Committee, chaired by Prof. Dr. Jameel-Un Nabi . The tremendous dedication of Dr. Zahid Halim and Dr. Shamsher Sadiq and all senior colleagues are also appreciated, whose engagement and endeavors to actualize this concept of the conference started a long time ago. We also wish to thank the organizers of 3rd Pak-Turk ETSE-2020, who followed cautiously the entire process of printing of this book of abstracts.

Editorial Team
Pak-Turk Conference,
GIK Institute, Topi, Pakistan.

IOP Conf. Series: Materials Science and Engineering **899** (2020) 011001 doi:10.1088/1757-899X/899/1/011001

3rd Pak-Turk International Conference on Emerging Technologies in the field of Sciences and Engineering 9-10th, June 2020

Aim and Scope of the Conference

The purpose of the conference was to provide a platform for researchers, academicians, and practitioners to become familiar with the latest advancements and technological developments in the various fields of sciences, technology, and engineering. The conference accepted a range of papers to encourage young and experienced researchers to present their work and the possibility of initiating mutual collaboration with international repute researchers and entrepreneurs of Pakistan and Turkey.

The scope of the conference includes, but is not limited to the following topics:

- Engineering Sciences
- Renewable Energy Resources
- Electrical Engineering
- Civil Engineering
- Data Science
- Data Mining
- Computational Intelligence
- Internet of Things
- Electronics
- Computer Science
- Material Science
- Chemical engineering
- Nuclear Astrophysics
- Optical Engineering
- Mechanical Engineering
- Solar cells
- Machine learning
- Intelligent Transportation Systems
- Virtual Reality
- Cloud Computing
- Intelligent Network Trends
- Vision and Image related trends
- Nanotechnology
- Mathematics

IOP Conf. Series: Materials Science and Engineering **899** (2020) 011001 doi:10.1088/1757-899X/899/1/011001

3rd Pak-Turk International Conference on Emerging Technologies in the field of Sciences and Engineering 9-10th, June 2020

Organizing Committees

PATRON-IN-CHIEF	**CONFERENCE STEERING COMMITTEE CHAIR**
Engr. Jehangir Bashar, Rector, GIK Institute	Prof. Dr. Jameel-Un Nabi, Pro-Rector (Academic)
Conference Secretary	**Conference Co-Secretary**
Dr. Zahid Halim Faculty of Computer Science and Engineering, GIK Institute	Dr. Shamsher Sadiq, Department of Civil Engineering, GIK Institute

IOP Conf. Series: Materials Science and Engineering **899** (2020) 011001 doi:10.1088/1757-899X/899/1/011001

3rd Pak-Turk International Conference on Emerging Technologies in the field of Sciences and Engineering 9-10th, June 2020

Editorial Board

IOP Conf. Series: Materials Science and Engineering **899** (2020) 011001 doi:10.1088/1757-899X/899/1/011001

3rd Pak-Turk International Conference on Emerging Technologies in the field of Sciences and Engineering 9-10th, June 2020

Program Schedule

9th June 2020 (Tuesday)		
Inauguration Program Schedule		
10:00-10:05 AM	Opening and Recitation of a few verses from the Holy Quran	
10:06-10:10 AM	National Anthems of Pakistan and Turkey	
10:11-10:15 AM	Theme and History of the Conference	Prof. Dr. Jameel-Un Nabi Pro-Rector (A) & Conference Steering Committee Chair
10:16-10:24 AM	Welcome Speech	Engr. Jehangir Bashar, Rector GIK Institute Conference Patron-In-Chief
10:25-10:35 AM	Address by the Chief Guest	Prof. Dr. Fazal Ahmad Khalid, SI
10:36-10:55 AM	Featured Keynote Talk1	Prof. Dr. Fazal Ahmad Khalid, SI
10:56-11:15 AM	Featured Technical Talk2 Stephen Hawking: Contribution to Humanity	Dr. Khalid Saifullah
11:16-11:35 AM	Featured Technical Talk3 AI for everyone	Dr. Yasar Ayaz
Keynote Addresses & Technical Talks		
Session Chair 1	**Prof. Dr. Jameel-Un Nabi**	
11:45-12:00 PM	Prediction of Geometric Shapes for even-even Ti, Cr, Fe Nuclei within the Interacting Boson Model (IBM) (Keynote)	Prof. Dr. Mahmut Boyukata
12:01-12:15 PM	Gravastars (Keynote)	Prof. Dr. Muhammad Sharif
12:16-12:30 PM	Optical Fiber Cavities with Ultrasensitive Detection Schemes for Liquid Sensing Applications	Dr. Imran Cheema
12:31 AM-12:45 PM	Nano-scale deposition of metal oxides and metal sulfides for multiple advance energy applications	Dr. M.A Basit
Lunch Break 12:45-01:15 PM		
Session Chair 2	**Prof. Dr. Mahmut Boyukata**	
01:16-01:30 PM	Statistical approach for hypernuclei formation at relativistic hadron and ion collisions (Keynote)	Prof. Dr. Nihal BUYUKCIZMECI
01:31-01:45 PM	Single Photon Quantum Mechanics (Keynote)	Prof. Dr. Muhammad Sabieh Anwar
01:46-02:00 PM	Enzyme function classification using deep learning	Dr. Hammad Naveed
02:01-02:15 PM	Summarizing big Graphs and recent trends	Dr. Kifayat Ullah Khan
Session Chair 3	**Prof. Dr. Muhammad Sharif**	
02:16-02:30 PM	Detecting fake twitter trends in an era of fake news	Dr. Syed Fawad Hussain

02:31-02:45 PM	An Automation System with Intelligent Production Scheduling and Cable Colors Detection for Cable Manufacturing in Automotive Industry	Dr. Doğan Aydın
02:46-03:00 PM	Technological developments for efficient quantum opto-electronic devices using molecular beam epitaxy (Industry)	Dr. Fauzia Jabeen
03:01-03:15 PM	Elliptic Curves: Geometry and Application	Dr. Umar Hayat
Session Chair 4	**Prof. Dr. Nihal BUYUKCIZMECI**	
03:16-03:45 PM	2D Materials: New Frontiers in Physics and Technology (Keynote)	Prof. Dr. Arshad Saleem Bhatti
03:46-04:00 PM	Investigation of Fermi Beta Transitions in Some Deformed Krypton (Kr) Isotopes	Dr. Necla Çakmak
04:00-04:15 PM	Beyond mean-field dynamics of quantum many-body systems	Dr. Bulent Yilmaz
04:36-05:15PM	**Parallel Session I: Physics and Material Science**	Session Chair(s) Dr. Fahd Nawaz
04:36-05:40 PM	**Parallel Session II: Computer Sciences & Engineering**	Session Chair(s) Dr. Ahmar Rashid
04:36- 05:15 PM	**Parallel Session III: Electrical Engineering**	Session Chair(s) Prof. Dr. Mohammad Akbar
06:00 -06:10 PM	**Parallel Session IV: Chemical Engineering**	Session Chair(s) Prof. Dr. Javaid Rabbani Khan
Paper Presentation 10-06-2020 (Day-2)		
09:00-11:20 AM	**Parallel Session I: Civil Engineering**	Session Chair(s) Prof. Dr. M. Ashraf Tanoli
09:00-12:10 PM	**Parallel Session II: Mechanical Engineering and Energy Systems**	Session Chair(s) Prof. Dr. S. M. Ahmad
Closing Program Schedule		
02:30-03:00 PM	The Physics Nobel Prize 2019 (Featured Keynote)	Prof. Dr. Asghar Qadir
03:01-03:30 PM	Laser Driven Proton Acceleration: An overview (Keynote)	Prof. Dr. Fida Yunus Khattak
03:31-03:40 PM	Recap and Conference Statistics	Dr. Zahid Halim Conference Secretary
03:41-03:45 PM	Concluding Remarks	Mr. Sardar Amin ullah Khan Pro-Rector (A&F)
03:45-03:51 PM	Message by Turkish Ambassador	H.E. Mr. Ihsan Mustafa Yurdakul
03:51-03:53 PM	Thought sharing by Turkish Speakers	Prof. Dr. Mahmut Boyukata
03:54-03:57 PM	Thought sharing by Turkish Speakers	Prof. Dr. Nihal BUYUKCIZMECI
03:58-04:02 PM	Thought sharing by Turkish Speakers	Dr. Necla Çakmak
04:03-04:06 PM	Thought sharing by Turkish Speakers	Dr. Doğan Aydın
04:07-04:10 PM	Thought sharing by Turkish Speakers	Dr. Bulent Yilmaz
04:10-04:20 PM	Best research papers (Third, Second, and First) announcement	Rector GIK Institute, Conference Patron-In-Chief
04:20-04:30 PM	Vote of Thanks	Rector GIK Institute, Conference Patron-In-Chief

IOP Conf. Series: Materials Science and Engineering **899** (2020) 011001 doi:10.1088/1757-899X/899/1/011001

3rd Pak-Turk International Conference on Emerging Technologies in the field of Sciences and Engineering 9-10th, June 2020

IOP Conf. Series: Materials Science and Engineering **899** (2020) 011001 doi:10.1088/1757-899X/899/1/011001

3rd Pak-Turk International Conference on Emerging Technologies in the field of Sciences and Engineering 9-10th, June 2020

IOP Conf. Series: Materials Science and Engineering **899** (2020) 011001 doi:10.1088/1757-899X/899/1/011001

IOP Conf. Series: Materials Science and Engineering **899** (2020) 011002 doi:10.1088/1757-899X/899/1/011002

Peer review declaration

All papers published in this volume of IOP Conference Series: Materials Science and Engineering have been peer reviewed through processes administered by the Editors. Reviews were conducted by expert referees to the professional and scientific standards expected of a proceedings journal published by IOP Publishing.

- **Type of peer review:** *Single-blind*
  Authors do not know about reviewers.

- **Conference submission management system:** *Online submission google form through conference website, https://www.giki.edu.pk/PakTurk/ImportantDates*

- **Number of submissions received:** *103*

- **Number of submissions sent for review:** *80*

- **Number of submissions accepted:** *48*

  - **Recommended for local proceedings:***30*
  - **Recommended for IOP MSE proceedings:***18*

- **Acceptance Rate (Number of Submissions Accepted / Number of Submissions Received X 100):**
  - **Overall conference:** *~46.6%*
  - **IOP Accepted***: ~17.5%*

- **Average number of reviews per paper:** *03*

- **Total number of reviewers involved:** *24*

- **Any additional info on review process:** *Reviewers were asked to reject or revise/accept based on standard review Performa. Revised/accepted papers were further categorised into local proceeding or IOP MSE proceeding recommendations based on quality of research.*

- **Contact person for queries:**

  Dr. Shamsher Sadiq , email: shamsher.sadiq@giki.edu.pk

IOP Conf. Series: Materials Science and Engineering **899** (2020) 012001 doi:10.1088/1757-899X/899/1/012001

Astrophysical S-factor for radiative capture of proton by ^{13}C at low energy

Abdul Kabir*[1], B. F. Irgaziev[1] and Jameel- Un Nabi[1]

[1]Faculty of Engineering Sciences, GIK Institute of Engineering Sciences and Technology, Topi 23640, Khyber Pakhtunkhwa, Pakistan.

*Email address: kabirkhanak1@gmail.com

Abstract. Radiative capture $p + {}^{13}\mathrm{C} \rightarrow {}^{14}\mathrm{N} + \gamma$ at energies of astrophysical interest is one of the important process in the CNO cycle. We focus the reader's attention on the possibility of describing this reaction within the framework of a single-particle potential model even when the reaction has a resonant characteristic. The partial components of the astrophysical S-factor are calculated for electric dipole transition. The calculated value of S-factor is in good agreement with experimental data both at low and high energy from the resonance position.

1. Introduction

Well known that the fusion reaction at low energy related to nuclear physics proceeds via formation the compound nucleus containing a large number of resonances. Therefore, the R-matrix approach is mostly used for description such reaction. The radiative capture reactions relevant to the nuclear astrophysics are going through both direct and resonant reactions at low energy. The description of the radiative capture data of the experiments mostly described by the R-matrix theory [1]. However, the number of resonances arising in collisions of a low-energy proton with light nuclei is few. Therefore, the resonant and non-resonant proton capture reaction can be described within the framework of a purely potential model. The solution by the potential model is good enough to yield cross sections within the accuracy required to reproduce the results of the experiments [2]. In stars massive than the Sun hydrogen-burning can proceed by the CNO cycle. The $^{13}\mathrm{C}(p, \gamma)^{14}\mathrm{N}$ is important for the calculation of the $^{12}\mathrm{C}/^{13}\mathrm{C}$ ratio, which is important to estimate the composition of different layers of the star. This reaction is seeding for $^{14}\mathrm{N}(p, \gamma)^{15}\mathrm{O}$ reaction which determines the rate of energy production in stars [3]. The ^{13}C is not only important astrophysically for the nucleo-synthesis but also as a source of the neutron in the intermediate-mass stars [4].

In this article, we calculate the S-factor for $^{13}\mathrm{C}(p, \gamma)^{14}\mathrm{N}$ capture from continuous state to ground state by using the potential model. The ground state wave function of ^{14}N was calculated by modifying the Woods-Saxon potential to achieve the correct behaviour of bound-state wave function in the ground state. Our calculation for the astrophysical S-factor reproduced the experimental results within the energy range of 0.01-1 MeV.

2. Formalism:

For calculation of astrophysical S-factor we take the total potential as

$$V(r) = V_N(r) + V_C(r), \qquad (1)$$

where in Eq. (1) $V_N(r)$ and $V_C(r)$ are the nuclear and Coulomb potentials respectively. The nuclear part of Eq. (1) we represent by the deformed Wood-Saxon potential to obtain the correct behaviour of the bound state wave functions. For calculation of the continuous state wave function, the deformation factor of the potential is selected as a unity.

$$V_n(r) = -\left[V_0 - V_{LS}(L \bullet S)\frac{1}{m_\pi r}\frac{d}{dr}\right]\frac{1}{1 + q(exp\frac{r - R_N}{a})} \quad (2)$$

where V_0, V_{LS}, and a in Eq. (2) represents the central potential depth, coupling strength of the spin-orbit potential and the diffuseness parameter, respectively; the product $(L{\bullet}S)$ is orbital and spin operators product; q is the deformation parameter. The nuclear radius R_N is defined by $R_N = r_0 \times A^{1/3}$, where A is mass number and r_0 is a parameter which varied within the range of 1.2–1.3 fm and m_π= 0.684 fm^{-1} is the pion mass in $\hbar = c = 1$ unit system. The potential defined by Eq. (2) describes correctly the asymptotic behaviour of the wave functions of both bound and continues states of the interacting nuclei. For the uniform charge distribution the Coulomb potential is

$$V_C = \begin{cases} \frac{hc}{2}\frac{\alpha Z_1 Z_2}{R_C}\left(3 - \frac{r^2}{{R_C}^2}\right) & If\, r \leq R_C, \\ \frac{hc}{2}\frac{\alpha Z_1 Z_2}{R_C} & If\, r \geq R_C, \end{cases} \quad (3)$$

where R_c is the Coulomb radius and α is the fine structure constant. For the p–^{13}C interaction the function $\phi_L(r)$ is the radial wave function which satisfies the radial part of the Schrodinger equation in both the initial and final states:

$$\frac{d^2}{d^2}\phi_L(r) + \frac{2\mu}{\hbar^2}\left[E - V(r) - \frac{\hbar^2 L(L+1)}{2\mu r^2}\right]\phi_L(r) = 0 \quad (4)$$

Where $V(r)$ is defined by Eq. (1) and E is the energy of relative motion of the interacting particles in the centre of mass (CM) system. The asymptotic behaviour of the continuous radial wave function is given by

$$\phi_L(r) \rightarrow Cos\delta_L F_L(kr) + Sin\delta_L G_L(kr) \quad (5)$$

Where k is the wavenumber corresponding to the interacting nuclei, δ_L is the elastic scattering phase shift, $F_L(kr)$ and $G_L(kr)$ are the regular and irregular Coulomb functions, respectively. The asymptotic of the radial bound state wave function is defined as

$$\phi_L(r) \rightarrow C_W W_{-\eta_0, L+\frac{1}{2}}(2k_0 r) \quad (6)$$

Where C_w is the single-particle asymptotic normalization coefficient [5]. The $W_{-\eta_0, L+\frac{1}{2}}(2k_0 r)$ represents the Whittaker function, L is the bound state orbital momentum, and η_0 is the Coulomb parameter $(\eta_0 = Z1Z2e^2\mu/k_0)$, k_0 is the wavenumber for bound state. For interaction of charged particles the astrophysical S-factor defines as

$$S(E) = \sigma(E)E\exp(2\pi\eta) \quad (7)$$

Where $S(E)$ and $\sigma(E)$ are the astrophysical S-factor and total cross section of the reaction, respectively; E is the collision energy in CM frame; η is Somerfield parameter defined as

$$\eta = Z_1 Z_2 e^2/\hbar v \qquad (8)$$

here Z_1 and Z_2 are the charge numbers of a projectile and target nuclei, respectively; v is their relative velocity; ħ is the Plank's constant. The total capture cross section is a sum over the multipolarity λ and the final state total angular momentum J_f

$$\sigma(E) = \sum_{J_f,\lambda} \sigma\lambda, J_f\,(E), \qquad (9)$$

where the summation term in Eq. (9) defined as

$$\begin{aligned}\sigma\lambda, J_f \;&= 8\pi\alpha \frac{C}{vk^2}[Z_1\left(\frac{A_2}{A}\right)^{\lambda} + Z_2(\frac{A_1}{A})^{\lambda}]^2 C^2 S_{J_f} \\ &\times \sum_{J_f,I,l_i} \frac{(k_\gamma)^{2\lambda+1}}{[(2\lambda+1)!!]^2} \frac{(\lambda+1)(2\lambda+1)}{\lambda} \\ &\times \frac{(2l_i+1)(2l_f+1)(2J_f+1)}{(2I_1+1)(2I_2+1)} \begin{pmatrix} l_f & \lambda & l_i \\ 0 & 0 & 0 \end{pmatrix}^2 \\ &\times \begin{Bmatrix} J_i & l_i & I \\ l_f & J_f & \lambda \end{Bmatrix}^2 (2J_f+1)\left(\int_0^\infty \phi_i(r) r^\lambda \phi_f(r) dr\right)^2 \end{aligned} \qquad (10)$$

here k_γ is the wave-number of the emitted photon, α is the fine structure constant, l_i and l_f are the orbital angular momenta of initial and final states, respectively. J_f, J_i are the total angular momenta of final and initial states, respectively, I_1 and I_2 are the spins of interacting nuclei, I is the channel spin, and k is the corresponding wave-number for the interacting particles in the initial channel. The $\phi_i(r)$ and $\phi_f(r)$ are the radial wave functions of initial and final states, respectively and r is the relative distance. The overlap integral which depends on the relative distance r is calculated by considering the initial and final states. Z_1, Z_2 and A_1, A_2 are the charges and mass numbers of the projectile and target particles, respectively. A is the total mass of interacting particles, C^2S_{Jf} spectroscopic factor, and $\begin{Bmatrix} J_i & l_i & I \\ l_f & J_f & \lambda \end{Bmatrix}$ are the 3j and 6j symbols, respectively [6-7]. Further, we restrict ourselves to calculations of electric dipole transitions, since they make the largest contribution to the cross section.

3. Results and Discussion:

We calculate the astrophysical S-factor for electric dipole transitions (E1) from ($J_\pi = {}^-1$, $T = 1$) to ground state within the range of energy of (0-1) MeV. For the calculation of cross section, it is necessary to calculate the bound state wave function. We calculate the bound state wave function of ^{14}N by modifying the Woods-Saxon potential to achieve better results of S-factor. Using Eq. (1) we calculate the bound state wave function for the ground state of ^{14}N by applying the parameters given in Table. (1). The same ground state wave function of ^{14}N for the ground state was calculated in [8] by using the Faddeev's method but they omitted the Coulomb potential. We compared our calculated bound state wave function with their calculated results in Fig. (1).

Table.1 the parameters of potentials equations. (2) and (3). The first column represents the state the rest columns represents the bound state energy measured, depth of central potential, spin-

IOP Conf. Series: Materials Science and Engineering **899** (2020) 012001 doi:10.1088/1757-899X/899/1/012001

orbit potential, diffuseness parameter, the nuclear and Coulomb radii the last column represents the deformation parameter in wood-Saxon potential.

State	E_f *(MeV)*	V_0 *(MeV)*	V_{LS} *(MeV)*	a *fm^{-1}*	R_N *fm*	R_C *fm*	q
Ground State							
$^{-}1$	-7.5505	102.74	14.0	0.5	0.286	3.056	4.0

For calculation of S-factor, we used Eq. (7). The capture of proton to resonance state ($J_\pi = {}^{-}1$, $T = 1$) can decay into the ground state in ^{14}N nucleus by E1 transition. The elastic scattering cross section for p–^{13}C contain the resonance at Er = 0.511 MeV. Here in our calculation, we take into account only E1 transitions because it is dominant within the energy range (0–1) MeV.

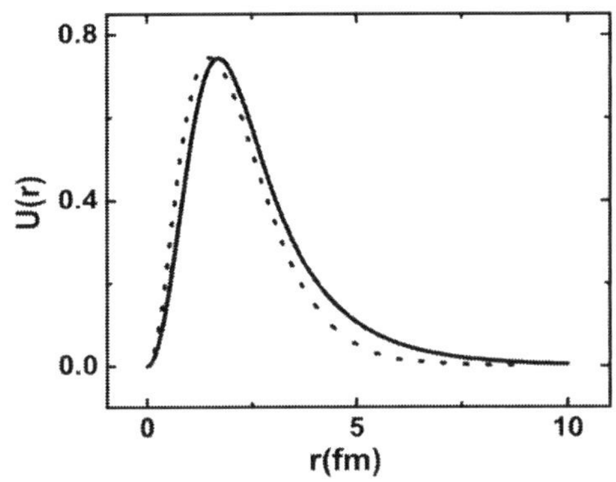

Figure 1 The solid line shows our calculations while the dotted line is the calculated results of [8].

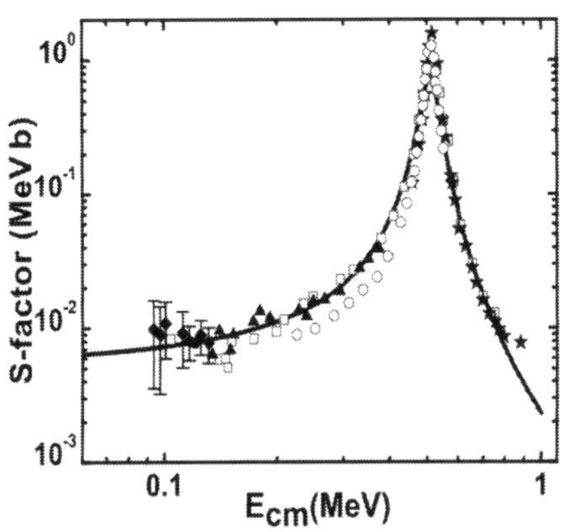

Figure 2. The solid line shows S-factor for ($J_\pi = {}^{-}1$, $T = 1$) to ground state. Filled lozenges [10], filled circle [11], filled triangles [12], open squares [9], open circles [13] and filled stars [14] represents the experimental data with error bars.

The results of the S-factor for this transition are shown in Fig. (2) and compared with experimental data. The values of spectroscopic factors for these transitions are 0.47. This value is obtained by the comparison of our calculated values of S-factor to the measured data of Ref. [9]. Our calculated S-factor is in good agreement in both low and higher energy.

4. Conclusion

In this article, we have used potential model for calculations of S-factor resonance transition. We modified the Woods-Saxon potential for bound state wave function. The modification is represented by q in The Woods-Saxon potential. We keep $q = 1$ for continuous state wave function. Based on this wave function we get the good results for the S-factor. So our modified parameters of Woods-Saxon potential correctly determine position of resonance. The calculated results for S-factor agree both at low and higher energy range from resonance peak.

References

[1] A. M. Mukhamedzhanov, et al. 2003 *Nucl. Phys. A.* **725** 279

[2] J. T. Huang, C. A. Bertulani, V. Guimaraes 2010 *At. Data and Nuclear Data Table.* **96** 824

[3] E. M. Burbidge, G. R. Burbidge, W. A. Fowler and F. Hoyle 1957 *Rev. Mod. Phys.* **29** 547

[4] M. Busso, R. Gallino, D. L. Lambert, C. Travaglio, V. V. Smith 2001 *J. Astrophys.* **557** 802

[5] G. R. Plattner, R. D. Viollier 1980 *Nucl. Phys A.* **365** 8

[6] R. F. Christy, I. Duck 1961 *Nucl. Phys.* **24(2)** 89

[7] B. F. Irgaziev, J-U Nabi, A. Kabir 2018 *Astrophys Space Sci.* **363** 8

[8] A. Moghadasi, H. Sadeghi and R. Pourimani 2018 *Astrophys Space Sci.* **2** 363

[9] J. D. King, R.E. Azuma, J.B.Vise, J. Gorres, C. Rolfs, H.P. Trautvetter, A.E. Vlieks 1994 *Nucl. Phys. A.* **567** 354

[10] R. E. Hester, W. A .S. Lamb 1961 *Phys. Rev.* **121** 584

[11] E. J. Woodbury, W. A. Fowler 1952 *Phys. Rev.* **85** 51

[12] D. F. Hebbard, J. L. Vogl 1960 *Nucl. Phys.* **21** 652

[13] G. Genard, P. Descouvemont, G. Terwagne 2010 *J. Phys. Conf. Ser.* **202** 012015

[14] V. J. Zeps, E. G. Adelberger, A. Garcıa, C. A. Gossett, H. E. Swanson, W. Haeberli, P. A. Quin, J. Sromicki 1995 *Phys. Rev C.* **51** 1494

An overview of methods for investigation of aeroelastic response on high aspect ratio fixed-winged aircraft

Muhammad Khizer Ali Khan[1], Ali Javed[1], Nafees Mumtaz Qadri[1], Mohtasshim Mansoor[1], and Farrukh Mazhar[1]

[1] Department of Aerospace Engineering, College of Aeronautical Engineering (CAE), *National University of Sciences and Technology (NUST),* Risalpur 24090, Pakistan

khizer_ali_khan@hotmail.com

Abstract. The thrust of increasing environmental and economic constraints on aircraft has enthused accelerated research in design of more economical and higher performance aircraft. Extensive experience in aerodynamic has established the use of high aspect ratio wings to improve the lift-to-drag ratio, a key parameter in determination of aircraft efficiency. Application of long slender wings has even more so intensified in the last decade due to emerging need for medium-to-high altitude long endurance (MALE/HALE) unmanned aerial vehicles (UAVs). However, such a wing comes at a trade-off of efficiency and safety. Longer wings tend of be more flexible and easily deform under load, hence are more vulnerable to the detrimental nature of aeroelastic effects. Divergence, control reversal and flutter are some major aeroelastic effects, which range from mere discomfort to complete destruction of body in flight. UAVs are most susceptible to this behaviour as their design incorporates very high aspect ratio wings. Numerous researches are available in literature which have focused on the explanation, calculation, and suppression of aeroelasticity; a subject which is as old as first heavier-than-air flight. This paper has attempted to cover the major aspect of aeroelasticity and summarize the state-of-the-art methods and approaches proposed by esteemed authors in this field for flutter prediction and suppression.
Keywords— Aeroelasticity; Aircraft; Divergence; Fixed-wing; Flutter; HALE; High-aspect ratio; MALE, UAV

1. Introduction

Aeroelasticity is an inter-disciplinary study of the combined effects of aerodynamic, elastic and inertial forces. It is the aeroelastic phenomena, which mostly puts a restrain on the flight envelop or more importantly, the maximum speed of flight for any vehicle (Figure 2). Aeroelastic effects can result in divergence, control reversal and flutter. Divergence is theoretically defined as the point where the aeroelastic twist of the structure becomes infinite and structure fails. If it is studied as a function of pressure, then aerodynamic pressure at which it occurs is known as the divergence pressure. At this point, the restoring forces of the elastic structure are not able to with-stand the aerodynamic loads and eventually gives away. In control reversal, the effectiveness of the control surfaces, such as aileron is reduced, and further reversed from its intended functionality. Flutter is a dynamic aeroelastic phenomenon, in which the aerodynamic, elastic and inertial forces interact to create a diverging vibrational response (Figure 1).

When the parameters which characterize the fluid-structure interaction (FSI) reach specific values where flutter occurs, they are called the critical values. In terms of physical phenomena, flutter is the point where "the energy of the flow is rapidly absorbed and transformed into the energy of mechanical vibrations" [2]. Flutter absorbs energy from the surroundings by the virtue of the fluid flow interaction over its surface that rapidly increases the amplitude of vibration.

Aeroelastic behavior of an aircraft has to be avoided either by the design of structure or use some active/ passive control mechanism that prevents or at least reduce the harmful effects. It is imperative to understand that the divergence, control reversal and flutter phenomena are an inherent property of structure and they cannot be eliminated. It can be observed even in many natural phenomena such as wings of birds, insects etc. The determination of its onset parameters such as speed on man-made machines helps to decide the conditions which should be avoided. Machines are designed such that aeroelastic effects initiate outside of the normal operating regime. Whole structure of aircraft deforms in flight due to the aerodynamic loads and its elastic nature, therefore, complete aircraft body should be treated as an elastic body rather than a rigid body [2].

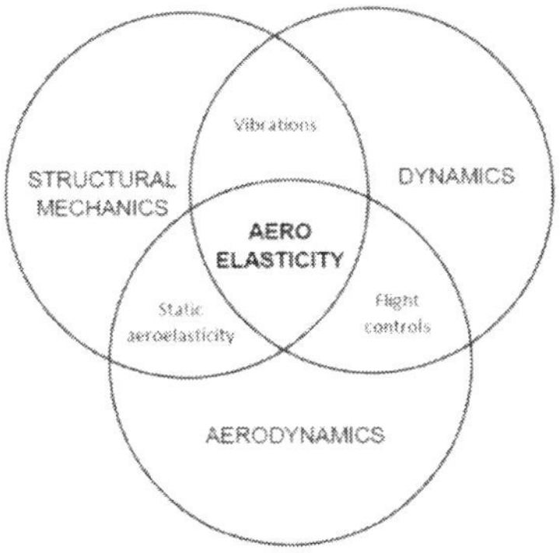

Figure 1. Venn diagram showing all possible relation between the set of disciplines that collectively forms the basis of aeroelasticity (adopted from Hodges) [1].

Although the effects of aeroelasticity are experienced over the whole body of an aircraft, its manifestation is exhibited most vibrantly by the wings and tails. These areas are subject to aeroelastic effects the earliest and they are most prone to fail under these effects. High aspect ratio (HAR) surfaces are more susceptible to this effect as they inhibit higher flexibilities and hence experience greater deflection under load. However, high-aspect ratio wings are important from the perspective of aerodynamic performance. Such wings are widely used in many aircraft as they give better fuel consumption and thus higher economy, with less environmental pollution. HAR wings reduces the aerodynamic drag, which accounts for almost half of the total drag of transport aircraft flying at cruise conditions [3]. This aerodynamic benefit is even greater at slower speed regimes [4] .

The concept of aeroelasticity is as old as the first aircraft, or even older than that [5]. Various tests such as Ground vibration tests (GVT), wind tunnel testing of dynamically-scaled model, theoretical analysis and flight flutter testing constitute an essential phase of aircraft design and development process. Researches aimed at devising improved models, approaches and method for such testing as well as techniques for suppression and control of aeroelastic effects are abundant in literature. It is still an active area essential for calculation, control and optimization of aeroelasticity, which is mandatory for certification and clearance of any aircraft, which is aspiring to take flight. All newly designed aircraft, or aircraft that goes through major modification in design or control system, or even has additional store attached to it, goes through this process of investigation.

Afonso et al. [6] discussed the various methodologies used to investigate the nonlinear aeroelastic effects which may arise due to nonlinearity in structure, aerodynamics and controls. Kehoe [7] presented a survey on the development of flight flutter testing techniques in which he discussed various in-flight excitation systems, instrumentation and data reduction methods which are used for the analysis of flight results. An overview of topics in aeroservoelasticity (ASE), computational and nonlinear aeroelasticity, rotary wing, and new technologies such as smart/adaptive structures are discussed by Friedmann [8]. Dowell et al. [9] provided a stability analysis on panel flutter, where plate and shell geometries have been discussed separately in addition to the analysis over theoretical and experimental discussion over the subject. Collar et al [10] provided with a brief insight on decade wise advances in the field for the first fifty years of the introduction of aeroelastic concept. Anderson et al [11] presented a review of latest techniques, analytical approaches and hardware used for the implementation of active aeroelastic control (AAC) in flutter suppression of aircraft. Livne [12] provided a 50 years summary of research and development in the area of active flutter suppression. When it comes to aeroelasticity, literature is abundant in collections that have provided valuable reviews and surveys of previous researches, challenges and future work to be carried out in this everlasting subject [13–18].

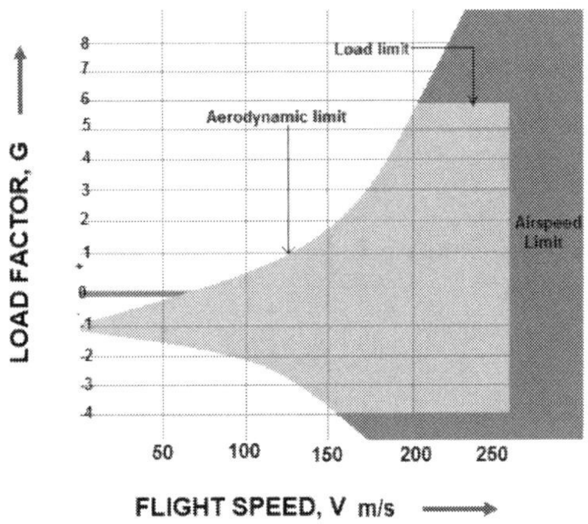

Figure 2. Figure shows a typical V-n diagram in which the flight speed (V) and load factor (G) boundaries are defined, that should not be exceeded to maintain structural integrity of aircraft.

The current survey paper briefly discusses the major aspects of aeroelasticity on aircraft surfaces, particularly high aspect ratio wings that are most vulnerable to the detrimental nature of such effects. This includes the various analytical, computational, and experimental approaches as well as flight testing methods used for investigation, calculation and optimization of divergence/ flutter boundaries that is of primary focus until present. Verification, validation, procedure-refining techniques on methodologies adopted to improve from previous mistakes and breakthroughs in research have been discussed. An analysis of methods proposed in the past for aeroelastic suppressing and expansion of flutter boundaries, which constitutes the subject of aeroservoelasticity has been summarized herein. The aim of paper is to extract reliable sources in the field of aeroelasticity that have provided novel methods and useful results. This would help readers better understand the phenomenon supported by results, and suggests the way forward for researchers interested in this area for a systematic and reliable study. Additionally, problems which continue to contest the advancement in fields of aeroelasticity and roadblocks to future developments have been discussed as challenges to this topic.

2. Analytical flutter analysis

Analytical flutter analysis can be regarded as the most powerful investigation tool as it allows for easy calculation of solution with variation of parameters. Hence it allows for a sensitivity study. Analytical analysis comes very early, whereas methods of experimental studies and flight-testing comes later in the design process. Therefore, analytically calculating the aeroelastic analysis of structure gives a good idea about the true behavior which is later evaluated using other approaches. The analytical process can be tailored in the later design phase when results from experiments and numerical analysis are available. Hence, after adjusting the analytical process, parametric and sensitivity studies can be carried out. The experimental setups are expensive and time consuming and can only be carried out for certain cases. Flight-testing, considered as the most dangerous of any test, comes latest in the design process.

Aeroelastic analysis can be divided into fully linear, quasi linear (linearized) and fully non-linear systems based on the linearity of three subsystems; structure, aerodynamics and control [5]. However, only structure and aerodynamic system will be discussed in this paper. The aeroelastic problem of fixed-wing aircraft is primarily the coupling of bending and torsion modes, which is essentially a linear problem [8]. Dowell et al [19] compared the experimental and theoretical results of a thin slender (high aspect ratio) wing using nonlinear uniform beam theory [20] and aerodynamic model form of ONERA. The effect of geometric nonlinearity on aerodynamic stall and flutter, that comes from the ratio of the ratio of flap and chord wise bending stiffness, EI_2/EI_1. The flutter stability boundaries are reduced for higher values of this ratio. Lee et al carried out investigation of the flutter boundaries on the 2-D airfoil placed in potential flow under the influence of cubic restoration structural force [21]. Wong et al [22] and Gong et al [23] have applied the methods developed for coupled nonlinear mechanical systems to predict the response on an airfoil with structural nonlinearities. Lee et al [24] used the Duffing's equation to model and solve the coupled 2 DoF system with cubic nonlinearities in both DoFs; bending and torsion.
Steady-state problems pose simple equations, like Am=n, where A is a square matrix. However, for equations such as dv/dt= Av, the solution is constantly changing with time, and hence cannot have a discreet value. The solution maybe simple sinusoidal, or continuously decaying or growing in nature. Hence, Eigen vectors or values are most significant in analyzing dynamic system. In eigenvalue framework, divergence occurs at the shifting of sign of real part of complex Eigen value. However, also the frequency at this point is 0 Hz. This is because divergence is a static phenomenon and it is not harmonic or oscillatory hence does not have any frequency associated with it. Flutter on the other hand is a dynamic phenomenon, which has a harmonic response that has an associated frequency with it. So in aeroelastic analysis, the sign-shifting of damping parameters with no frequency is the divergence point, and point with some associated frequency is the flutter point (Fig 3).

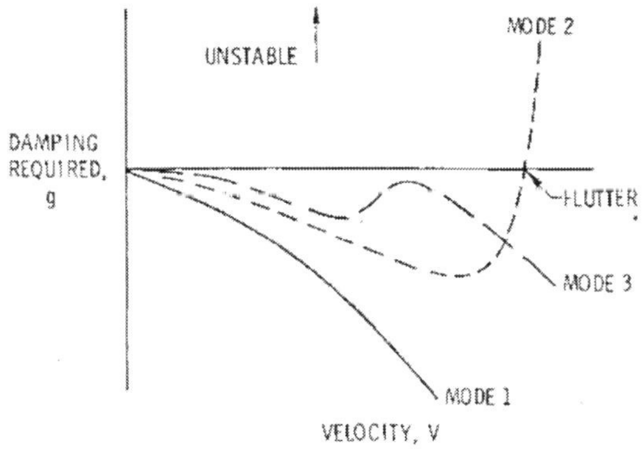

Figure 3. V-g method of flutter analysis shows the modal damping for second mode shifts in the unstable region which marks the flutter point [25].

Not much emphasis has been given in the past to the optimization of flutter boundaries of structure, which is evident from the lack of research available in this area. Keeping some aspects constrained such as mass, area or geometry of the structure, the optimal values for the remaining parameters can be calculated to obtain the best possible objective function. Objective function in the case of flutter is dynamic pressure which is meant to increase. Particle swarm optimization is an effective approach to find the optimal solution of a problem by iteratively evaluating a problem and grading based on a given property or quality of system. It works in the same way as a swarm of insects. It was first proposed by Kennedy and Eberhert [26] and since then has been used by many research for optimal solution of their given problem. Researcher have applied this approach in the solution of their aeroelastic analysis to obtain parameters which would increase the aerodynamic pressure of aircraft, which is usually the single most significant design objective to achieve. Solution optimization of should be given its due significance as it provides good result with little or no trade-off.

Torabi et al. [27] carried out optimization of flutter boundaries for a trapezoidal plate made of FGM sandwich plates. They used first order shear deformation theory and piston theory for the modelling of FGM sandwich plates and aerodynamic pressure respectively. FGM plates consists of an isotropic interior metal core, sandwiched between FG sheets that has an Al2O3 ceramic rich exterior and Al rich metal interior. Transition between the materials is carried out using a power law function. Differential quadrature method (DQM) was used to solve the governing equations and boundary conditions for trapezoidal-shaped cantilevered FG sandwich plate in supersonic regime. Their result were optimized using particle swarm optimization (PSO), which yields the best values of plate design parameters. Their study proved that the adopted approach yields results with acceptable accuracy and fast convergence rate. This type of constrained optimization is quiet useful for designing of wings and tail of aircraft [28, 29].

3. Computational Aeroelasticity

Software that use computational methods for the solution of fluid flow are termed as computational fluid dynamic (CFD) and for structural analysis are called computational structural dynamics (CSD). Packages that couple these two tools for aeroelastic analysis are called Computational Aeroelasticity (CAE) software [30]. However, it is injustice to constrain CAE to such simplistic definition considering the extensive research which has led to its current state.

3.1 Fidelity of models

Advancement in computational capability has led to the development of higher fidelity models. However, the type of models to be selected depends on factors such as the complexity of structure, accuracy requirement, computational time, and available funding for the project. Low-to-medium fidelity models are used at the initial levels of analysis for quicker results which help to make bigger decisions in the design process. Once the design process is finished, refined and higher fidelity models are employed for capturing of detailed aeroelastic picture which is not adequately addressed in low fidelity models. However, using discretization methods (FE) with potential flow aerodynamics in inviscid, incompressible and low compressible region is ideally used in medium fidelity models. For aerodynamic modelling, either 2-D strip theory, 3-D lifting surfaces theories or 3-D panel theories (incorporates thickness) maybe used depending on the requirement [31]. Doublet lattice Method (DLM) is considered most popular for flutter prediction [32, 33]. Higher fidelity CFD models are restricted to Euler (inviscid) and Navier Stokes equations (viscous) [34, 35].

3.2 Meshes

Meshing of geometries is carried out either using structured or unstructured methods. Structured mesh offer better computational efficiency and does not require a connectivity matrix for solution. Unstructured meshes are easier to use for complex configurations as it reduces the human and computational effort [30]. Researchers are expected to have sufficient knowledge about mesh refinement and areas where it needs to be incorporated for capturing of relevant phenomena. FSI packages are designed using adaptive algorithms, which try to adjust according to requirement. Still some input and care by the user is required to ensure accuracy of results.

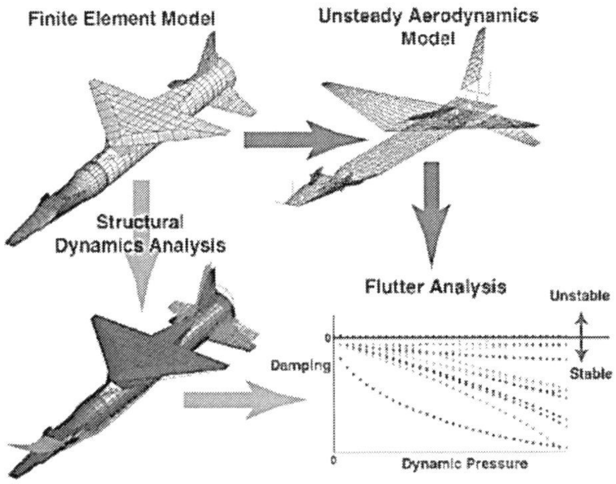

Figure 4. An example of structural and aerodynamic modelling of launch vehicle for computational flutter analysis is shown [36].

4. Experimental analysis

A single approach to research in an area is never considered reliable unless the researcher has some reference values from which the result of that study can be compared. A validation in the approach and methodologies adopted needs to be carried out to convince the readers about the validity of the results. In the case of aircraft, it becomes even more so important because a single approach, no matter how rigorous, is not dependable in any area. The next step in aeroelastic investigation of aircraft is experimental study, in which reduced models are used for wind tunnel experiments or flight flutter testing (or both). The scaling is done to adjust the size of model inside the test section of wind-tunnel. Ground vibration test is done prior to wind-tunnel experiment to determine the modal behaviour of scaled down model. These tests provide a valuable insight in the behaviour of structure, which otherwise is left unexplained in numerical study. The experimental results not only provide answers in a short span of time, but also assist in adjusting and aligning the development of its theory and numerical model.

4.1 Aeroelastic scaling

Downscaling of life-sized model is carried out to accommodate the model inside the limited space of test sections as wind-tunnel experiment of full-scale model is not feasible. Sizing has to be done with regards to scaling laws so that the behavior of the actual model can be reproduced. Beside researches, wind tunnel tests are also carried out for testing of modifications, testing of active-control, and mandatory for flutter clearance requirement. Classical scaling techniques have been compiled by Bisplinghoff [37] for linear aeroelasticity, however, literature is not very abundant in research for nonlinear scaling methods. Recently some work has been done in this area due to the increasing use of high aspect ratio wings [38]. Bond et al. [39] was able to match the first natural frequencies of first three modes, their mode shapes and damping ratio for non-linear aeroelastic scaling. Cesnik et al [40] used nonlinear stiffness matrix to account for the geometric nonlinearities and concluded that the Froude number should be catered for scaling structures. Spada et al [41] presented an updated version on Ricciardi et al [42] and compared his results with classical scaling techniques to prove their methodology gave better flutter speeds, frequencies and other modal results. For cases where flutter response is of primary interest, one should ensure that the rigid and dynamic modes remain same for reduced and full scale model.

4.2 Ground vibration tests (GVTs)

Before moving on to conducting wind-tunnel experiment of flight testing of aeroelastically reduced scaled models, it is recommended to first carry out model updation and validation through ground vibration tests. Research on the GVT methodologies on high aspect ratio wings is scarce in literature as the dynamic response for high aspect ratio wing aircraft is difficult to analysis as compared to a conventional aircraft. The method has become even more difficult with the introduction of complex design in structures and use composite materials in the inner structure of wing [43, 44]. NASA has played a key role in analysis of such cases under project PAAW (Performance adaptive Aeroelastic Wing) in which GVT of a flying wing aircraft is presented [45]. While many researches have focused on assessing the presence of non-linearity, very few have actually suggested ways in which it could be addressed [46].

4.3 Wind Tunnel Experiments

Wind tunnels provide a closed environment for a controlled study of wing. Some safety aspects need to be covered to ensure the safety of model as well as the wind tunnel. Results from wind tunnel helps to validate the numerical and theoretical models. Reliable extraction of data from wind tunnel requires carefully calibrated mechanisms, and a sensitive measuring, and excitation device. A choice of model support system for mounting of model also needs to be considered to effectively capture aeroelastic trends. Sting mount and side-turn table was suggested by Ricketts et al [47] as the simplest form of support. Traditional wind tunnel tests in which the aircraft is constrained are not suitable for a high aspect ratio wing aircraft SU et al. [48] as it does not allow for the coupling of rigid and dynamic structural modes. Keeping in view, a support that caters for the flexibility of the structure is proposed by Tang et al. [49].

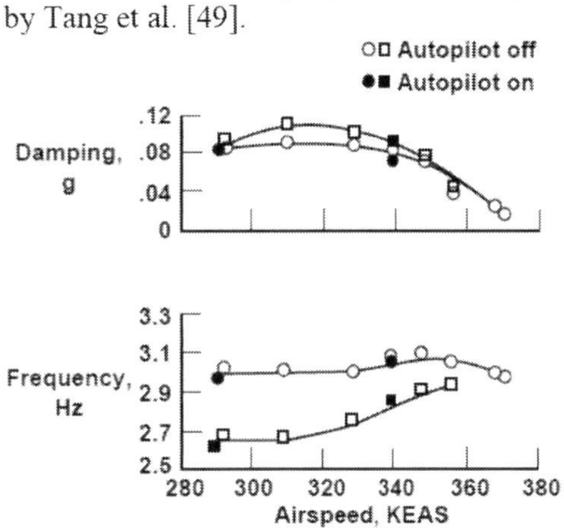

Figure 5. An example of structural and aerodynamic modelling of launch vehicle for computational flutter analysis is shown [36].

4.4 Flight Flutter Tests

For commercial airliners, it is the requirement of aviation regulation authorities to prove through flight-testing that aircraft is free from aeroelastic effects well beyond the normal flying envelop of aircraft so that safety of vehicle as well as passenger(s) is ensured. UAVs and drones are also flight tested for flutter to ensure airworthiness of wings and testing of any applied controlling technique. It is imperative to carry out flutter test on new aircraft, or aircraft that has undergone structural and/or controls modification. Even the installation of additional external stores on aircraft such as drop tanks or missile necessitates the need for flight-testing. Flutter flight test is one of the most dangerous in aviation. Data is collected through various sensors such as accelerometers attached on aircraft. The structure is excited for its vibration modes. Using data, the damping characteristics and how it changes with flight conditions is evaluated [7, 50]. There is a damping response of each mode, and some modes are highly damped whereas some modes are not. The point where the mode damping becomes equivalent to the structural damping is the flutter point for that mode, and beyond this point, amplitude of self-excited oscillation continues to increase and aircraft enters into flutter. The point where this damping occurs at the lowest aerodynamic pressure is obviously the most important and the restricting factor of maximum aircraft speed.

For flight flutter test, pilots are briefed to attain the targeted speed, and keep going until flutter is experienced. Actually, the purpose is not to enter the flutter domain, but come close to it as it only takes a matter of seconds for a structure to completely disintegrate if it flies past the flutter boundary. Therefore, the pilot or controller in case of UAV always approach flutter speed and sense the points where they start to feel discomfort in the aircraft behavior. The speed of aircraft at which this indication occurs is noted and is regarded as the never exceed speed. The normal operating range of aircraft is set well below this point. Results of flight flutter testing are rarely report in literature. Kehoe [7] presented a survey on the development of flight flutter testing techniques in which he discusses various in-flight excitation systems, instrumentation and data reduction methods which are used for the analysis of flight results.

5. Active Flutter suppression

Traditional passive flutter suppression techniques such as mass balancing and local stiffening are inefficient as they add to the overall weight of structure [51]. Perhaps the only remedy to reduce the detrimental nature of flutter without any design modifications in the long span wing of MALE/HALE UAV is the introduction of an actively controlled mechanism in the system. Extensive research in this area has been carried out in the recent years, where a suitable feedback system continuously monitors some parameters on the vehicle and accordingly adjusts an actively controlled mechanism that prevents or delays the onset of flutter. An example of such system would be the movement of control surfaces according to a control law, which relates the motion of the surfaces with the dynamic pressure over the wing. The forces are modified in such a way that the structural turbulence is alleviated and hence the safety margin is increased. Recent flight test demonstrations, wing tunnel tests and analytical works have shown promising results.

Implementation of hydro-servo systems for active control techniques used in surface deflections such as flaps, aileron, elevens etc. are accompanied with problem of response time. Hydraulic actuator movement is sluggish, and not quick enough to immediately address flutter onset or gust alleviation. Hence, such a system cannot cope with high frequency oscillations. Lu et al. [52, 53] have contributed in the study of thin airfoil flutter suppression using active acoustic excitation. Acoustic devices have the advantage of simple hardware implementation, large frequency range and most importantly the very quick response which is essential in dealing with high frequencies on oscillating body. They [53] have shown in paper that a simple loudspeaker can be appropriately adjusted to stabilize an airfoil experiencing flutter. The incident acoustic wave interacts with the trailing edge to produce favourable vortex shedding that suppresses the flutter motion. The location of source field, gains constants, and phase of gain constants plays in significant role in the acoustic control design.

The thrust of research in smart structures has led to the used to such materials in giving continuous structural deformation of the wings and other lifting surfaces which can be used to alter the unsteady aerodynamics loads and suppress flutter. Several studies on the use of piezoelectric materials in static aeroelasticity as well as suppression of flutter using piezoelectric actuation are available in literature. Much of its potential is demonstrated through wind-tunnel tests for incompressible flow which are available in literature.

Friedmann et al [54] have shown using piezoelectric actuation, increase of dynamic pressure by as much as 12%, hence improving the flutter margin. Despite having a substantial potential of piezoelectric actuators in aeroservoelasticity, not much research on the scaling of such actuators have been carried out which would relate the behavior of actual wing with that a different sized models. Moniz et al [55] devised a novel scaling technique for flow in compressible regimes in which feedback control using piezoelectric actuators, or conventional trailing edge flaps are used to control and extend the flutter margin.

6. Morphing aircraft

All aircraft ever made are subject to a range of flight conditions such as during climb, descent, cruise, maneuver, and landing/taking off. A wing that operates optimally at all these varying speeds and environments does not exist. Therefore, aircraft designers use wing that operates most efficiently in the dominant part of the vehicles flight and operates sub-optimally at other regimes. Small and low aspect ratio wings allow for higher speeds and better maneuverability but have poor aerodynamic performance. High aspect ratio wings such as those used on MALE/HALE UAVs may offer good range and efficiency, but lack maneuverability and are prone of flutter at higher speeds. Advancements in materials, structural concepts, methods of actuation, and MIMO (many input many output) system has motivated research and development of variable-shape vehicles; the morphing vehicles [14, 56, 57]. Multidisciplinary optimization (MDO) is obtained through the integration of smooth shape variation, major planform change, smart actuation and active flow control system to obtain desired flow patterns in areas of geometric and flow discontinuity. These features distinguishes the morphing airplane from conventional shape-changing aircraft, which considers limited shape-variations in sweep, camber, and dihedral angles. A novel concept for a simple wing morphing technique was proposed by Gamboa et al. [58]. in which the span of a composite wing was made to change in flight. Wing consisted of an inboard part, housing smaller, moveable outboard moving part that is slid using servomotor. It was shown that the safety margin was flight speed was improved by 44% A gear driven autonomous twin spar (GNATSpar) was used by Ajaj et al. [59] in which inner space of opposite wing is used to house the unused length of the spar. Spar extends and retracts based on velocity feedback that in turn changes the wingspan. Lift to drag ratio of the wing was increased by 30% without any aeroelastic penalty.

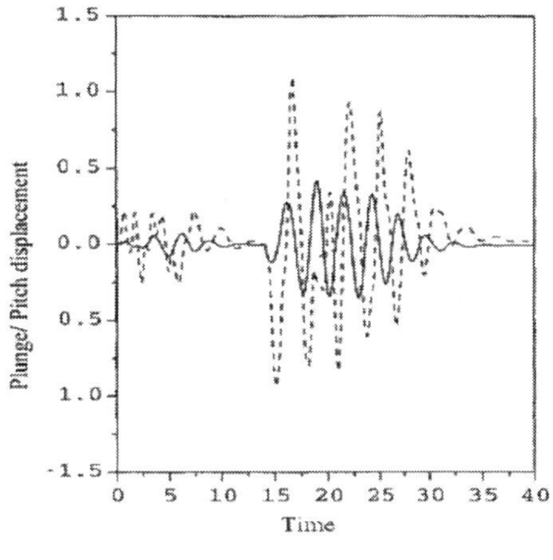

Figure 6. The time response of pitch and plunge displacement of an airfoil in flutter is shown with and without the application of active suppression system (bold and dotted line respectively) [8].

Though a promising technology, manufacturers are still skeptical of benefits offered by morphing mechanism to adopt in their design [60] . Many interesting concepts for application of morphing technology in airplane have been proposed, but very few have actually been proven through wing tunnel testing and even fewer have flown. Morphing wing system is also accompanied by an additional weight penalty, which must be overcome.

7. Conclusion

Credit for the creation of engineering marvels likes the F-16 and Concorde commercial aircraft goes to the generations of aeroelasticians who have brought aeroelasticity to its current state. Surely, the biggest constrain in the desire for higher speeds and more economy comes from the grave nature of aeroelastic effects. Any contribution in the solution for the challenges posed by this field is considered a breakthrough. Sincere efforts must also be put in to preserve the work of past challenges, their breakthroughs and lesson which come from extensive experience. It is indeed a rich and complex multidisciplinary field and it is difficult to cover the details of every area within the confines of this paper. However, the present paper is an educational effort that is aimed to compliment recent surveys with minimum overlap. Topics of various methods employed for the study, calculation and suppression of aeroelastic response in general and flutter in particular have been considered. The targeted audience of this survey are novice researchers in this field, which intends to allow them to grasp the basic idea of this phenomena as well as identify the gaps, which would lead readers in direction of active research. This survey paper has effectively discussed topics of analytical, computational, and experimental techniques, aeroservoelasticity and aeroelastic design optimization of high aspect ratio wings. Paper contains reference to further surveys and researches, which should be read for an extensive insight on a particular subject. Aeroelasticity is an extensive field, not bound by the limited areas touched upon in this paper. Despite its history, topic of aeroelasticity is far from saturation, and there exists a wide area in all headings highlighted in this paper, where a great potential for research exists.

8. Recommendation

In most flutter investigation, the immobility of fuselage is considered which neglects any rolling motion of the plane, so that the wing motion is entire due to its structural deformation. The inclusion of fuselage mobility could be included in future research, which accompanies an additional degree of freedom. Although the complexity of the problem would be increased, study could be carried out to determine the extent to which the results are improved. A research gap is observed in the GVT of scaled models in which research for checking the presence of non-linearity in high aspect ratio wings is available, however methods to address those non-linearity are scarce in literature. Active control system such as piezoelectric actuation provide a novel solution for suppression of static and dynamic aeroelastic effect on wings. Validation of its effect on scaled model through wind tunnel testing requires scaling. No significant work has been reported for the aeroelastic and aeroservoelastic scaling since the 1960s. Future research could be focused on suggesting ways in which the scaling of actuation systems could be carried out so that the response on full-scale models could be predicted. Active research in smart materials that utilizes the advancements in material and structures technology, aerodynamics, sensing, actuation and control system for multidisciplinary design optimization (MDO) should be encouraged. Morphing wings allow for major shape variation throughout the flight envelop which is well beyond the old, conventional variable-shape vehicles that holds great potential in drastic flutter suppression and prevention, which is most imminent for application on UAVs.

References

[1] D. H. Hodges and G. A. Pierce, *Introduction to structural dynamics and aeroelasticity*: Cambridge University Press, 2011.
[2] M. A. Shubov, "Flutter Phenomenon in Aeroelasticity and Its Mathematical Analysis," *J. Aerosp. Eng.*, vol. 19, no. 1, pp. 1–12, 2006.
[3] A. Abbas, J. de Vicente, and E. Valero, "Aerodynamic technologies to improve aircraft performance," *Aerospace Science and Technology*, vol. 28, no. 1, pp. 100–132, 2013.
[4] E. Torenbeek, *Advanced aircraft design: Conceptual design, analysis and optimization of subsonic civil airplanes*. Hoboken, New Jersey: John Wiley & Sons, 2013.
[5] E. H. Dowell, Ed., *A modern course in aeroelasticity*. Cham: Springer, 2015.
[6] F. Afonso, J. Vale, E. Oliveira, F. Lau, and A. Suleman, "A review on non-linear aeroelasticity of high aspect-ratio wings," *Progress in Aerospace Sciences*, vol. 89, pp. 40–57, 2017.
[7] M. W. Kehoe, "A historical overview of flight flutter testing," 1995.
[8] P. P. Friedmann, "Renaissance of Aeroelasticity and Its Future," *Journal of Aircraft*, vol. 36, no. 1, pp. 105–121, 1999.
[9] E. H. Dowell, "Panel flutter - A review of the aeroelastic stability of plates and shells," *AIAA Journal*, vol. 8, no. 3, pp. 385–399, 1970.

[10] A. R. Collar, "The first fifty years of aeroelasticity," *Royal Aeronautical Society Journal (AIAA)*, vol. 5, pp. 12–20, 1978.
[11] *International Mechanical Engineering Congress and Exposition*: American Society of Mechanical Engineers, 2004.
[12] E. Livne, "Aircraft Active Flutter Suppression: State of the Art and Technology Maturation Needs," *Journal of Aircraft*, vol. 55, no. 1, pp. 410–452, 2018.
[13] R. M. Bennett and J. W. Edwards, "An overview of recent developments in computational aeroelasticity," in *29th AIAA Fluid Dynamics Conference*, Albuquerque, USA, 1998, pp. 1–14.
[14] E. Livne, "Future of Airplane Aeroelasticity," *Journal of Aircraft*, vol. 40, no. 6, pp. 1066–1092, 2003.
[15] J. Njuguna, "Flutter prediction, suppression and control in aircraft composite wings as a design prerequisite: a survey," *Struct. Control Health Monit.*, vol. 14, no. 5, pp. 715–758, 2007.
[16] Y.C. Fung, *An introduction to the theory of Aeroelasticity*. New York, 1993.
[17] Jan R. Wright and Jonathan E. Cooper, *Introduction to Aircraft Aeroelasticity and Loads*. Sussex, England: John Wiley & Sons, 2007.
[18] Martin Hollmann, *Modern Aerodynamic flutter analysis*. Monterery, California, 1991.
[19] D. Tang and E. H. Dowell, "Experimental and Theoretical Study on Aeroelastic Response of High-Aspect-Ratio Wings," *AIAA Journal*, vol. 39, no. 8, pp. 1430–1441, 2001.
[20] D.H. Hodges and E. H. Dowell, *"Nonlinear Equations of Motion for the Elastic Bending and Torsion of Twisted Nonuniform Rotor Blades,"*: NASA TN D-7818, 1974.
[21] B. H. K. Lee, L. Y. Tiang, and Y. S. Wong, "Flutter of airfoil with cubic restoration force," *J. Fluids Struct*, vol. 13, pp. 75–101, 1999.
[22] Y. S. Wong, B. H. K. Lee, and L. Gong, "Dynamic response of a two-degree of freedom system with cubic nonlinearity," in *Proc. 3rd Int Conf. on computational physics*, 1995.
[23] L. Gong, Y. S. Wong, and B. H. K. Lee, "Dynamics of a coupled system of Duffing's equations," *Dynamics of Continuous, Discrete and Impulsive Systems*, vol. 4, pp. 99–119, 1998.
[24] Lee and B. H. K. et al., "Analysis and Computation of Nonlinear Dynamic Response of a Two-degree-of-freedom System and its Application in Aeroelasticity," *Journal of Fluids and Structures*, no. 11, pp. 225–246, 1997.
[25] I. E. Garrick and W. H. Reed, "Historical Development of Aircraft Flutter," *Journal of Aircraft*, vol. 18, no. 11, pp. 897–912, 1981.
[26] R. Eberhart and J. Kennedy, Eds., *Particle swarm optimization*, 1995.
[27] K. Torabi and H. Afshari, "Optimization for flutter boundaries of cantilevered trapezoidal thick plates," *J Braz. Soc. Mech. Sci. Eng.*, vol. 39, no. 5, pp. 1545–1561, 2017.
[28] J. Li and Y. Narita, "Analysis and optimal design for supersonic composite laminated plate," *Composite Structures*, vol. 101, pp. 35–46, 2013.
[29] J. Li and Y. Narita, "Multi-objective design for aeroelastic flutter of laminated shallow shells under variable flow angles," *Composite Structures*, vol. 101, pp. 35–46, 2014.
[30] D. M. Schuster, D. D. Liy, and L. J. Huttsell, "Computational aeroelasticity: success, progress, challenges," *Journal of Aircraft*, vol. 40, no. 5, pp. 843–856, 2003.
[31] K.M. Kier, "Comparison of unsteady aerodynamic modeling methodologies with respect to flight loads analysis," in *Proceedings of the AIAA Atmospheric Flight Mechanics Conference*, California, USA: AIAA, 2005, Paper 90-0978.
[32] *MSC Software, Tech. Rep. MD Nastran 2011 and MSC Nastran 2011 release guide, http://www.mscsoftware.com/, 2011.*
[33] *ZONA technology, Inc., Tech. Rep. ZAERO Version 8.3, User's manual, Scottsdale, AZ, USA, (http://zonatech.com/ZAERO.htm).*

[34] J.D. Anderson, *Fundamentals of Aerodynamics*. New York: Mc-Graw Hill, 2001.
[35] J. Moran, *An Introduction to Theoretical and Computational Aerodynamics*. Newyork, USA: John Wiley & Sons, 1984.
[36] D. M. Schuster, D. D. Liu, and L. J. Huttsell, "Computational Aeroelasticity: Success, Progress, Challenge," *Journal of Aircraft*, vol. 40, no. 5, pp. 843–856, 2003.
[37] R. L. Bisplinghoff, H. Ashley, and R. L. Halfman, *Aeroelasticity*. Newyork, USA: Dover Publications, 1996.
[38] Z. Wan and C. E. Cesnik, Eds., *Geometrically Nonlinear Aeroelastic Scaling for Very Flexible Aircraft*. Boston, Massachusetts: AIAA, 2013.
[39] V. L. Bond, R. A. Canfield, A. Suleman, and M. Blair, "Aeroelastic Scaling of a Joined Wing for Nonlinear Geometric Stiffness," *AIAA Journal*, vol. 50, no. 3, pp. 513–522, 2012.
[40] Z. Wan and C. E. S. Cesnik, "Geometrically Nonlinear Aeroelastic Scaling for Very Flexible Aircraft," *AIAA Journal*, vol. 52, no. 10, pp. 2251–2260, 2014.
[41] A. Suleman, F. Afonso, and C. Spada, Eds., *AIAA SciTech: 54th AIAA Aerospace Sciences Meeting*, 2016.
[42] A. P. Ricciardi, R. A. Canfield, M. J. Patil, and N. Lindsley, "Nonlinear Aeroelastic Scaled-Model Design," *Journal of Aircraft*, vol. 53, no. 1, pp. 20–32, 2016.
[43] M. W. Kehoe and L. C. Freudinger, "Aircraft ground vibration testing at the NASA dryden flight research facility," NASA, TM-104275, 1993.
[44] M. W. Kehoe, "Aircraft ground vibration testing at NASA Ames-Dryden Flight Research Facility," NASA, TM-88272, 1987.
[45] A. Gupta, P. J. Seiler, and B. P. Danowsky, Eds., *Ground Vibration Tests on a Flexible Flying Wing Aircraft- Invited: AIAA Atmospheric Flight Mechanics Conference*, 2016.
[46] C. P. Moreno, A. Gupta, H. Pfifer, B. Taylor, and G. J. Balas, Eds., *Structural model identification of a small flexible aircraft*. San Diego, California, USA,: AIAA, 2016.
[47] R.H Ricketts, "Experimental aeroelasticity – history, status and future in brief," in *Proceedings of the 31st AIAA/ASME/ASCE/AHS/ASC Structures, Structural Dynamics, and Materials Conference*, Long Beach, California, USA: AIAA, 1990, Paper 90-0978.
[48] W. Su and C. E. Cesnik, "Nonlinear aeroelasticity of a very flexible blended-wing-body aircraft," *Journal of Aircraft*, vol. 47, no. 5, pp. 1539–1553, 2010.
[49] D. Tang and E.H. Dowell, "Experimental and theoretical study on gust response for high-aspect-ratio wings," *AIAA Journal*, vol. 40, no. 3, pp. 419–429, 2002.
[50] A. A. Abbasi and J. E. Cooper, "Statistical evaluation of flutter boundaries from flight flutter test data," *Aeronaut. j.*, vol. 113, no. 1139, pp. 41–51, 2009.
[51] Y. H. Zhao, "Flutter suppression of a high aspect-ratio wing with multiple control surfaces," *Journal of Sound and Vibration*, vol. 324, no. 3-5, pp. 490–513, 2009.
[52] P. J. Lu and L. J. Huang, "Optimal control law synthesis for flutter suppression using active acoustic excitations," *Journal of guidance, control, and dynamics*, vol. 16, no. 1, pp. 124–131, 1993.
[53] P. J. Lu and L. J. Huang, "Flutter suppression of thin airfoils using active acoustic excitations," *AIAA Journal*, vol. 30, no. 12, pp. 2873–2881, 1992.
[54] P. P. Friedmann and E. Presente, "Active control of flutter in compressible flow and its aeroelastic scaling," *Journal of guidance, control, and dynamics*, vol. 24, no. 1, pp. 167–175, 2001.
[55] A. Suleman and P. Moniz, "Active aeroelastic aircraft structures," in *Proc. III European Conf. on Computational Mechanics*, Lisbon, Portugal, 2006, pp. 85–106.

[56] R. G. Loewy, "Recent Developments in Smart Structures with Aeronautical Applications," *Journal of Smart Materials and Structures*, vol. 5, pp. 11–41, 1997.
[57] C. Thill, J. Etches, I. Bond, K. Potter, and P. Weaver, "Morphing skins," *Aeronaut. j.*, vol. 112, no. 1129, pp. 117–139, 2008.
[58] P. Gamboa, P. Santos, and J. Silva, "Flutter analysis of a composite variable-span wing," in *4th International Conference on Integrity, Reliability and Failure*, Funchal, 2013, pp. 23–27.
[59] R. M. Ajaj, M. I. Friswell, M. Bourchak, and W. Harasani, "Span morphing using the GNATSpar wing," *Aerospace Science and Technology*, vol. 53, pp. 38–46, 2016.
[60] S. Barbarino, O. Bilgen, R. M. Ajaj, M. I. Friswell, and D. J. Inman, "A Review of Morphing Aircraft," *Journal of Intelligent Material Systems and Structures*, vol. 22, no. 9, pp. 823–877, 2011.

IOP Conf. Series: Materials Science and Engineering **899** (2020) 012003 doi:10.1088/1757-899X/899/1/012003

Implementation of Differential Transform Method (DTM) for Large Deformation Analysis of Cantilever Beam

Hassaan Abbasi and Dr. Ali Javed
Aerospace Engineering Department, National University of Sciences & Technology (NUST), College of Aeronautical Engineering (CAE), Risalpur, Pakistan

h.abbasi@cae.nust.edu.pk; hassaan16612@gmail.com

Abstract. Large deformation of a cantilever beam is an eminent structural analysis and engineering problem. A non-linear differential equation governs the problem of large deformation of beams. It has many important applications scientific and engineering fields. It is quite intricate and complex to find a closed-form or an exact solution for such a problem. It is relatively easier to use analytic expressions for calculations than numerical or experimental analysis. Large deformation analysis has always been an active area of research for researchers. The main focus of this research is to implement a new method namely Differential Transformation Method (DTM) for non-linear beam problem analysis. DTM is quite reliable and efficient in giving an approximate analytical solution to beam problems. Moreover, it is a new application for DTM. Hence, the primary purpose of this work is to evaluate the utility of the method for solution of beam problem. DTM yields a 2-point boundary value problem. It is based Taylor's series expansion. DTM helps to calculate the deflection angle and horizontal as well as vertical displacements of a cantilever beam, experiencing large deformation, in an explicit analytical form. Mathematical formulation code will be developed using MATLAB. Variation of rotation angle at free end will be analysed using different number of terms used and values of non-dimensional load "β". Results of DTM prove the method to be quite effective and suitable for predicting solutions to non-linear beam problems. Therefore, DTM can be used for widespread applications for new, emerging and complex engineering problems.

Index Terms— Cantilever beam, Large deformation, Differential Transformation method (DTM).

1. Introduction

Mechanical systems which involve non-linearity due to large deflection of the members continue to be a thought-provoking problem for engineers. As a result, quite many research works have been stated on the issue of the non-linear large deformation of beams. Chebyshev's polynomial method was used for large deformation analysis by Schmidt and Dadeppo [1]. Large deformation of a spring-hinged beam was investigated by Nageswara Rao et al. [2] the with a tip concentrated load. Nageswara Rao and Venkateswara Rao [3] studied the large deformations of a non-uniform beam with a rotational load at end. Wang et al. [4] investigated large deformation beam problems in which one end of the beam was able to slide freely while the other end was fixed. The beam with a point load was solved using Elliptic integral method and the shooting-optimization technique.

Lee [5] studied large deformation of Ludwick made cantilever beams with a combined loading effect of a concentrated load at tip and uniformly distributed load over length of beam. Chucheepsakul and Phungpaigram [6] studied closed-form solutions with the help of elliptic integrals for large deformation of an elastic beam. Dado and Al- Sadder [7] studied prismatic as well as non-prismatic beams for very large deformations for various loading conditions. It was based on representation of rotation angle by a polynomial of the position along the deformed axis of beam. Integral of the residual was minimized and beam boundary conditions were applied for determining polynomial co-efficients. Wang et al. [8] employed Homotopy Analysis Method (HAM) for analysis of large deflection of a cantilever beam subjected to point load. Tolou and Herder [9] studied the feasibility of the Adomian decomposition

IOP Conf. Series: Materials Science and Engineering **899** (2020) 012003 doi:10.1088/1757-899X/899/1/012003

method (ADM) for analysis of complex mechanical systems. Mutyalarao et al. [10] studied a uniform cantilever beam for large deformation, subjected to a concentrated load at tip and having an inclination normal to the deflected beam axis.

It is often easier to use analytic expressions for calculations than to employ experimental or numerical techniques. It is also a clear preliminary point for a better comprehension of the association between the physical parameters of beam as well as slope at any point along the beam arc length. Besides, the quest for an analytical solution for the complex non-linear differential equation of large deformation of beams subjected to a concentrated load is of great scientific and engineering concentration. The key resolve of this research work is to establish the efficacy of the DTM for solution of the above-mentioned complex problem. Large deformation of beams is a novel application for the DTM and was previously implemented for other engineering problems.

2. Problem Statement

The curvature ρ of a beam with uniform cross-section is given as: -

$$\rho = \frac{d\theta}{ds} = \frac{M}{EI} \tag{1}$$

where:
M: Flexural moment
E: Young's or Elastic modulus
I: Beam moment of inertia
EI: Beam Flexural stiffness
θ(s): slope with respect to the horizontal axis
s: coordinate of arc along beam neutral axis

As shown in ***Fig1***, a beam is under deflection due to a point load at tip. Moment at any point in beam is given as:-

$$M = F(L - \delta_h - x) \tag{2}$$

F being concentrated load at tip. Hence, equation of a prismatic beam for large deformation, is: -

$$\frac{d\theta}{ds} = \frac{F}{EI}(L - \delta_h - x)\ ;\ \theta(0) = 0\ ;\ \theta'(L) = 0 \tag{3}$$

δ_h being horizontal deflection of beam. Since the lateral deflection is much larger than axial elongation at tip, therefore, axial elongation of the beam has been neglected. Differentiation of equation (1) gives:-

$$\frac{d^2\theta}{ds^2} = \frac{\frac{dM}{ds}}{EI} \tag{4}$$

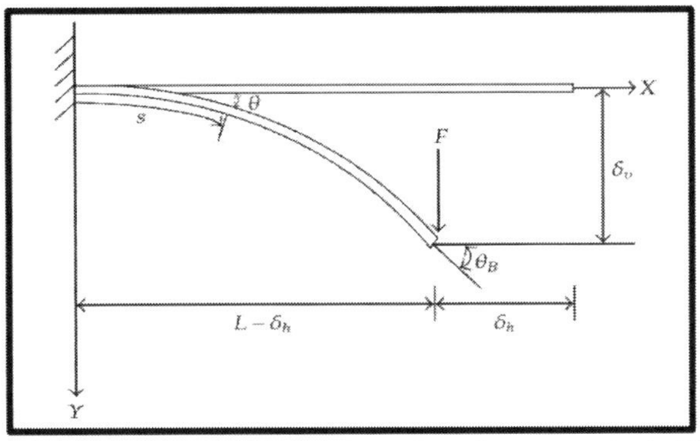

Figure 1: Large deformation of a beam under tip concentrated load [30, 31].

Let $\boldsymbol{\eta} = \boldsymbol{s/L}$ be a dimensionless parameter for position on arc length of beam. Differentiation of equation (2) with respect to *s*, and substituting in equation (4), following form of governing equation and boundary conditions for large deflection of beam is yielded:-

$$\frac{d^2\theta}{d\eta^2} + \beta cos\theta = 0 \qquad (5a)$$
$$\theta(0) = 0\,; \qquad (5b)$$
$$\theta'(1) = 0 \qquad (5c)$$

Where $\beta = FL^2/EI$ being the tip point load in dimensionless form. $\Theta_{tip} = \theta(1)$ is the beam rotation angle at free end. Non-dimensional horizontal displacement $\boldsymbol{\delta_h}$ of tip is given by [10, 30, 31] :-

$$\frac{L-\delta_h}{L} = \sqrt{\frac{2EI(sin\theta_B)}{FL^2}} = \sqrt{\frac{2sin\theta_B}{\beta}} \qquad (6)$$

$$\frac{\delta_h}{L} = 1 - \sqrt{\frac{2sin\theta_B}{\beta}} \qquad (7)$$

Equation (7) shows that non-dimensional horizontal displacement $\boldsymbol{\delta_h}$ depends upon rotation angle at tip $\boldsymbol{\theta_B}$ and non-dimensional end point load $\boldsymbol{\beta}$. For an infinitesimal deformation, equation (5a) may be written as:-

$$\frac{d^2\theta}{d\eta^2} + \beta = 0\,;\ \theta(0) = 0;\ \theta'(1) = 0 \qquad (8)$$

And the solution is given as: -

$$\theta(\eta) = \frac{\beta}{2}(2-\eta)\eta \qquad (9)$$

For tip $\boldsymbol{\eta} = \mathbf{1}$ and: -

$$\theta_B = \frac{\beta}{2} \qquad (10)$$

In order to consider large deformation, a non-linear equation has to be solved [10]:

IOP Conf. Series: Materials Science and Engineering **899** (2020) 012003 doi:10.1088/1757-899X/899/1/012003

$$\sqrt{\beta} = K(\mu) - F(\varphi, \mu) \qquad (11)$$

Where:

$$\mu = \sqrt{\frac{1 + sin\theta_B}{2}} \quad ; \quad \varphi = arcsin\left(\frac{1}{\sqrt{2}\mu}\right) \qquad (12)$$

$K(\mu)$ and $F(\varphi, \mu)$ being complete elliptic integral of the first kind and elliptic integral of the first kind respectively.

3. Analysis of methods

3.1. Differential Transformation Method (DTM) [32]

Differential transformation of a function $x(t)$ is given by [33]:

$$X(k) = \frac{(H)^k}{k!}\left[\frac{d^k x(t)}{dt^k}\right]_{t=0} \qquad (13)$$

where
x(t): Function to be transformed
X(k): Transformed function

Similarly, inverse transform of *X(k)* is given by: -

$$x(t) = \sum_{k=0}^{\infty}\left(\frac{t}{H}\right)^k X(k) \qquad (14)$$

Differential transformation is actually founded upon expansion of Taylor series. As the number of discretes (values of *X(k))* is increased, it becomes more precisely possible to restore the function to be determined. The function *x(t)* can be represented as a finite series and equation (18) can be re-written as:

$$x(t) = \sum_{k=0}^{N}\left(\frac{t}{H}\right)^k X(k) \qquad (15)$$

Basic mathematical operations of DTM are listed in table in ***figure 2***.

Original function	Transformed function
$x(t) = \alpha f(t) \pm \beta g(t)$	$X(k) = \alpha F(k) \pm \beta G(k)$
$x(t) = \frac{df(t)}{dt}$	$X(k) = (k+1)F(k+1)$
$x(t) = \frac{d^2 f(t)}{dt^2}$	$X(k) = (k+1)(k+2)F(k+2)$
$x(t) = t^m$	$X(k) = \delta(k-m) = \begin{cases} 1 & k = m \\ 0 & k \neq m \end{cases}$
$x(t) = \exp(\lambda t)$	$X(k) = \frac{\lambda^k}{k!}$
$x(t) = f(t)g(t)$	$X(k) = \sum_{l=0}^{k} F(l)G(k-l)$

Figure 2: Fundamental operations of DTM [32].

IOP Conf. Series: Materials Science and Engineering **899** (2020) 012003 doi:10.1088/1757-899X/899/1/012003

3.1.1. Solution with DTM

Applying differential transformation to equation (5a) with respect to η, and taking $H=1$, gives: -

$$(k+2)(k+1)\Theta(k+2)-\frac{1}{2}\beta\sum_{l=0}^{k}\Theta(l)\,\Theta(k-l)+\frac{1}{24}\beta\left(\sum_{m=0}^{k}\Theta(k-m)\left(\sum_{v=0}^{m}\Theta(m-v)\left(\sum_{w=0}^{v}\Theta(v-w)\,\Theta(v)\right)\right)\right)+\beta\times\delta(k)=0 \qquad (16)$$

Applying transformation on boundary conditions, we get: -

$$\Theta(0)=0 \qquad (17)$$

$$\Theta(1)=C \qquad (18)$$

where constant C is determined using the boundary condition in equation (5c) for point $\eta=1$. Using, inverse differential transformation, $\Theta(k+2)$ is calculated from equation (20) as follows: -

$$\Theta\left(\frac{1}{2}\right)=-0.5\beta \qquad (19a)$$

$$\Theta(3)=0 \qquad (19b)$$

$$\Theta(24)=\frac{1}{24}\beta C^2 \qquad (19c)$$

$$\Theta(5)=-\frac{1}{40}\beta^2 C \qquad (19d)$$

$$\Theta(6)=\frac{1}{240}\beta^3-\frac{1}{720}\beta C^4 \qquad (19e)$$

$$\Theta(7)=\frac{1}{336}\beta^2 C^3 \qquad (19f)$$

$$\Theta(8)=-\frac{13}{6720}\beta^3 C^2 \qquad (19g)$$

$$\Theta(9)=\frac{1}{1920}\beta^4 C-\frac{1}{8640}\beta^2 C^5 \qquad (19h)$$

Substituting equations (19 a-h) into equation (15), the closed-form solution is given as: -

$$\theta(\eta)=C+\Theta(2)\eta^2+\Theta(4)\eta^4+\Theta(5)\eta^5+\Theta(6)\eta^6+\Theta(7)\eta^7+\Theta(8)\eta^8+\Theta(9)\eta^9+\Theta(10)\eta^{10}+\Theta(11)\eta^{11}+\Theta(12)\eta^{12}\ldots\ldots\ldots\ldots \qquad (20)$$

Substitution of the boundary condition from equation (5c) into above equation, in order to obtain the value of C, yields: -

$$\theta'(1)=0=-\beta+\frac{1}{6}\beta C^2-\frac{1}{8}\beta^2 C+\left(\frac{1}{40}\beta^3-\frac{1}{120}\beta C^4\right)+\frac{7}{336}\beta^2 C^3-\frac{13}{840}\beta^3 C^2+\cdots=0 \qquad (21)$$

Now $\theta(\eta)$ can be determined by substitution of C into equation (20). N=12 was found to be enough to provide an accurate solution. Therefore, equation (5) involves only a limited number of terms and the solution can be calculated without unnecessary computational effort.

4. Results and discussion

Using equations of DTM described above, deflection angle at tip θ_{tip} and dimensionless horizontal displacement δ_h/L were found by varying dimensionless point load β and number of terms N. Calculations were made using MATLAB (code appended at the end of paper).

Table 1 shows convergence of θ_{tip} for different values of non-dimensional point load β by increasing number of terms used for transformation. Bold figures indicate the values, which do not change up to fourth place by increasing N any further. For $\beta=0.5$, value of θ_{tip} doesn't change after N=8, whereas for other greater values of α, value doesn't change upto fourth place decimal after $N=12$.

Table 1: θ_{tip} vs N for different values of β.

N	ϑ_{tip}			
	β=0.5	β=1.0	β=1.5	β=2.0
4	0.249479	0.495833	0.735937	0.966666
6	0.249479	0.495833	0.735937	0.966666
8	**0.24948**	0.495885	0.736333	0.968333
10	0.24948	0.495885	0.736333	0.968333
12	0.24948	**0.495884**	**0.736321**	**0.968247**
14	0.24948	0.495884	0.736321	0.968247
16	0.24948	0.495884	0.736321	0.968247

Table 2 shows convergence of δ_h/L for different values of non-dimensional point load **β** by increasing number of terms. Since δ_h/L depends on θ_{tip} and **β**, it follows same trend as does θ_{tip}.

Table 2: δ_h/L vs N for different values of β.

N	δ_h/L			
	β=0.5	β=1.0	β=1.5	β=2.0
4	0.006221	0.024537	0.053933	0.092808
6	0.006221	0.024537	0.053933	0.092808
8	**0.006219**	0.02449	0.053726	0.092287
10	0.006219	0.02449	0.053726	0.092287
12	0.006219	**0.024491**	**0.053733**	**0.092314**
14	0.006219	0.024491	0.053733	0.092314
16	0.006219	0.024491	0.053733	0.092314

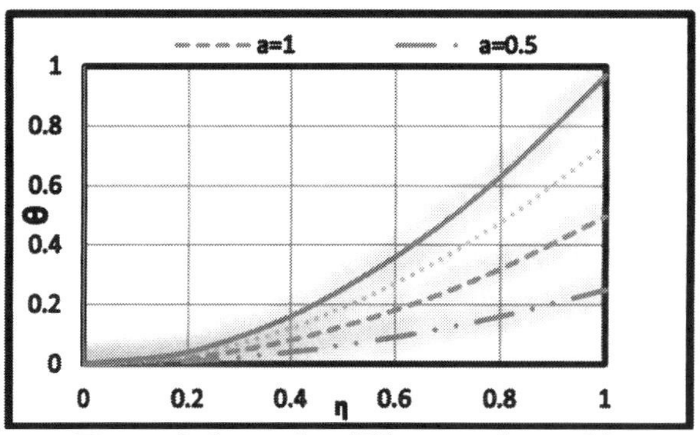

Figure 3: θ vs η for different values of β.

Figure 3 illustrates variation of deflection angle θ at different cross-section locations of beam η for various point load values of β. It is clearly depicted that increasing β as well as η, increases deflection angle at a particular cross-section. Furthermore, as β is increased beyond 1, increase in θ becomes more and more significant for a particular value of η. For values of β less than 1, increase in deflection angle is not that much substantial.

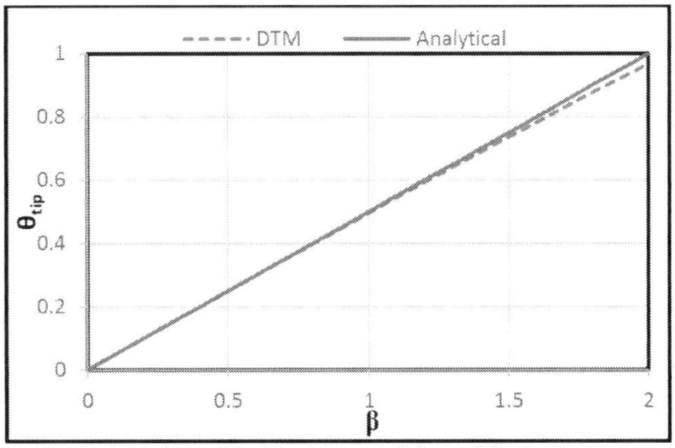

Figure 4: Variation of θ_{tip} with β.

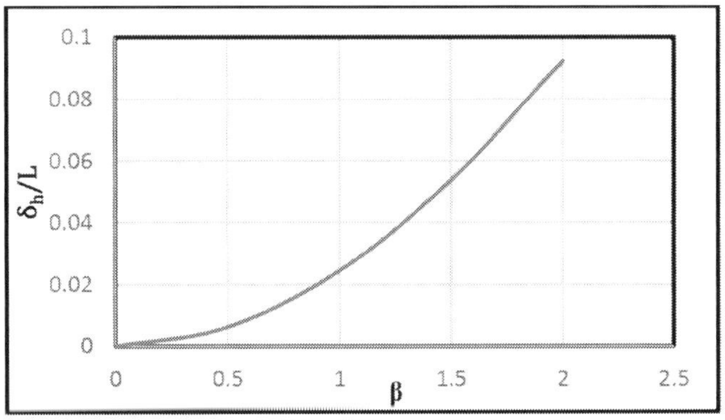

Figure 5: Variation of δ_h/L vs β.

Figures 4 and 5 illustrate variation of tip deflection angle θ and non-dimensional tip horizontal displacement δ_h/L with dimensionless point load β, respectively. Figure 4 compares the results obtained from DTM and actual analytical results. It can be seen that DTM provides results with quite great accuracy. For values of β greater than 1, difference between two values increases, as it is evident from Eq. (24) that effect of β will be greater for larger values of β and also by increasing number of terms N. Figure 5 shows that for $\beta=2$, tip horizontal displacement is almost one-tenth of the total beam length. Also, for $\beta>1$, variation of δ_h/L with β is linear.

the paper is optional and left as a decision for the author. Where the author wishes to divide the paper into sections the formatting shown in table 2 should be used.

5. Conclusion

DTM provides a reliable approximate explicit analytical solution to large deformation of cantilever beam under point load. Results obtained from DTM are quite accurate up-to fourth decimal place. Deflection angle, horizontal displacement as well as vertical displacement of beam cross-sections can be determined with the help of DTM. Hence, DTM offers an accurate approximate solution non-linear beam problem considering large deformation. DTM may also be applied to other such non-linear complex engineering problems to determine explicit approximate solutions.

6. Recommendations

Following are the recommendations for further research and study:

(a) Efficacy of DTM for other boundary conditions and loading conditions may be studies in order to determine its accuracy for implementation in those cases.

(b) Results may be compared with other methods like Variational Iteration Method (VIM) for same boundary and loading conditions.

7. Appendix A

Appended below is the MATLAB code used for calculations using DTM:

```
%Code for Differential Transformation Method for large Deformation
analysis
%of a cantilever beam
clc
clear all
close all
syms a c z
t1=c;
t2=-0.5*a*z^2;
t3=0;
t4=1/24*a*c^2*z^4;
t5=-1/40*a^2*c*z^5;
t6=((1/240*a^3)-(1/720*a*c^4))*z^6;
t7=1/336*a^2*c^3*z^7;
t8=-13/6720*a^3*c^2*z^8;
t9=((1/1920*a^4*c)-(1/8640*a^2*c^5))*z^9;
t10=((257/1209600*a^3*c^4)-(1/19200*a^5))*z^10;
t11=((-67/443520*a^4*c^3)+(1/475200*a^2*c^7))*z^11;
t12=((157/2956800*a^5*c^2)-(13/1182720*a^3*c^6))*z^12;
t13=((529/31449600*a^4*c^5)-(7/748800*a^6*c))*z^13;
t14=((-
120383/9686476800*a^5*c^4)+(1/3548160*a^3*c^8)+(1/1497600*a^7))*z^14
;
t15=((547/106444800*a^6*c^3)-(391/399168000*a^4*c^7))*z^15;
t16=((25469/18598035456*a^5*c^6)-(15107/12300288000*a^7*c^2)-
(19/5474304000*a^3*c^10))*z^16;
t17=((-
13752547/13173608448000*a^6*c^5)+(4337/130288435200*a^4*c^9)+(163/10
18368000*a^8*c))*z^17;
ThetaZ=[t1;t2;t3;t4;t5;t6;t7;t8;t9;t10;t11;t12;t13;t14;t15;t16;t17];
prompt='Please enter N(must be greater than or equal to 4 and less
than or equal to 16';
N=input(prompt)
ThetaZ1=sum(ThetaZ(1:N),1)
prompt='Please enter value of alpha (Non-dimensional load
parameter)';
a=input(prompt)
prompt='Please enter value of Eeta (Non-dimensional position along
beam';
z=input(prompt)
c=0;%real values of c are very very small and can be neglected
ThetaZ1=subs(ThetaZ1)
```

8. References

[1] R. Schmidt and D. A. Dadeppo, Approximate analysis of large deflections of beams, Zeitschrift für Angewandte Mathematik und Mechanik, 51 (1971) 233-234.

[2] B. Nageswara Rao and G. Venkateswara Rao, On the large deflection of cantilever beams with end rotational load, Zeitschrift für Angewandte Mathematik und Mechanik, 66 (10) (1986) 507-509.

[3] B. Nageswara Rao and G. Venkateswara Rao, Large deflections of a cantilever beam subjected to a rotational distributed loading, Forschung im Ingenieurwesen, 55 (4) (1989) 116-120.

[4] B. Nageswara Rao, G. L. Nagesha Babu and G. Venkateswara Rao, Large deflection analysis of a spring hinged cantilever beam subjected to a tip concentrated rational load, Zeitschrift für Angewandte Mathematik und Mechanik, 67 (10) (1987) 519-520.

[5] B. Nageswara Rao and G. Venkateswara Rao, Large deflections of a nonuniform cantilever beam with end rotational load, Forschung im Ingenieurwesen, 54 (1) (1988) 24-26.

[6] C. M. Wang, K. Y. Lam, X. Q. He and S. Chucheepsakul, Large deflection of an end supported beam subjected to a point load, International Journal of Non-Linear Mechanics, 32 (1997) 63-72.

[7] K. Lee, Large defections of cantilever beams of non-linear elastic material under a combined loading, International Journal of Non-Linear Mechanics, 37 (2002) 439-443.

[8] S. Chucheepsakul and B. Phungpaigram, Elliptic integral solutions of variable-arc-length elastic under an inclined follower force, Zeitschrift für Angewandte Mathematik und Mechanik, 84 (1) (2004) 29-38.

[9] M. Dado and S. Al-Sadder, A new technique for large deflection analysis of non-prismatic cantilever beams, Mechanics Research Communications, 32 (2005) 692-703.

[10] J. Wang, J. K. Chen and S. Liao, An explicit solution of the large deformation of a cantilever beam under point load at the free tip, Journal of Computational and Applied Mathematics, 212 (2008) 320-330.

[11] N. Tolou and J. L. Herder, A semi-analytical approach to large deflections in compliant beams under point load, Mathematical Problems in Engineering, Article ID 910896, 13 pages (2009) doi:10.1155/2009/910896.

[12] M. Mutyalarao, D. Bharathi and B. Nageswara Rao, Large deflections of a cantilever beam under an inclined end load, Applied Mathematics and Computation, 217 (2010) 3607-3613.

[13] A. A. Joneidi, D. D. Ganji and M. Babaelahi, Differential transformation method to determine fin efficiency of convective straight fins with temperature dependent thermal conductivity, International Communications in Heat and Mass Transfer, 36 (2009) 757-762.

[14] H. Yaghoobi, P. Khoshnevisrad and A. Fereidoon, Application of the differential transformation method to a modified van der pol oscillator, Nonlinear Science Letters A, 2 (4) (2011) 171-180.

[15] H. Yaghoobi and M. Torabi, The application of differential transformation method to nonlinear equations arising in heat transfer, International Communications in Heat and Mass Transfer, 38 (2011) 815-820.

[16] J. M. Gere and S. P. Timoshenko, Mechanics of materials, PWS, Boston, Mass, USA (1997).

[17] J. K. Zhou, Differential transform and its applications for electrical circuits, Wuhan, Huarjung University Press (1986).

[18] A. H. Hassan, Differential transformation technique for solving higher-order initial value problems, Applied Mathematics and Computation, 154 (2004) 299-311.

[19] Salehi, P., Yaghoobi, H. & Torabi, M., Application of the differential transformation method and variational iteration method to large deformation of cantilever beams under point load, J Mech Sci Technol 26 (9) (2012) 2879~2887.

IOP Conf. Series: Materials Science and Engineering **899** (2020) 012004 doi:10.1088/1757-899X/899/1/012004

On Board Intelligence for Public Transportation in Developing Countries using IoT

A Rehman[1] **and T Mumtaz**[2]

[1] Data Communication & Control (Pvt) Ltd, 409 - 410, Clifton Centre, Block 5 Clifton, Karachi 75600, Sindh, Pakistan
[2] Habib University, Block 18, Gulistan-e-Jauhar, Off Shahrah-e-Faisal Rd, Faisal Cantonment, Karachi, Sindh, Pakistan

engr.arehman1995@gmail.com, tariq.mumtaz@habib.edu.pk

Abstract. This study is carried out for indigenous development of an ITS (Intelligent Transportation System) device aimed particularly for public transport buses in the developing countries with poor infrastructure and unaccountable transportation services. The developed system uses IoT technology, extendable hardware and open source software in order to achieve commercial scalability by government and transportation authorities in such countries. The proposed device essentially collects and store data from the bus fleet and communicates the information to a central server from which software applications for passengers and transport operators are developed accordingly. The device functionality is divided into three categories which are Environment Sensing, Vehicle Management and Safety. The acquired dataset can be used to produce useful insights that can benefit the social and commercial sector of the city. Analytics such as fuel consumption, vehicle health, predicting driver behaviour, forecasting of rush-hours and vehicle maintenance requirements can be estimated. The device is packaged into a box with proper considerations of reliability in terms of power, weather proofing, storage and cabling. The proposed device will allow the passengers, especially in developing cities, to plan their journeys ahead of time and provide a holistic decision support system to the transport authorities.

1. Introduction

Transportation is one of the most pressing issues in developing countries like Pakistan. This study is situated at the confluence of issues such as passenger accessibility to transportation infrastructure, access to science and technology, design thinking, and rights and entitlement. This low-cost prototype is designed to address the issue of lack of dependable public transportation and infrastructure in developing countries, a case which is reasonably complex due to social, political and economic challenges in such countries. This study aims to propose and investigate new methods of improvement in public transportation of developing countries where some form of transportation mechanism exists.

This paper proposes an indigenously developed ITS device, aiming to build a sustainable model that shall encourage the citizens to reduce the usage of private vehicles and start using the public transportation. Inspired by commercial vehicle telematics devices, the proposed Onboard intelligent unit is designed with features of Fleet Management/ Automatic Vehicle Location System (AVL) and Real Time Passenger Information (RTPI). This paper is structured as follows: Section II provides a brief overview of similar devices and solutions developed globally. Section III presents the proposed system model and its characteristic features. Section IV describes in detail about the hardware and software model along with its design and interface for various users. Finally, Section V discusses the conclusion and on-going future work.

2. Related Survey

Different solutions are practiced at different parts of the world. A brief analysis of the products and services used in public buses as part of ITS, in the past decade, includes: TMS (Transportation Management System) [1], Real-Time Bus Information System [2], APTS (Advanced Public Transportation System) [3] and DVR based systems. These systems take input from using smart sensing and cyber-physical systems and output collected information using various M2M (Machine-to-Machine) or IoT protocols. Some alternatives focus on the passenger safety and information availability, while others focus on improving the quality and accessibility of public transit systems. In essence, most of the recent advances in ITS devices focus on allowing the passengers and bus operators to take advantage of the data and provide users with the necessary information to enhance the transit service quality.

In the past decade, the IoT based Gateway systems have gained significant commercial attention of technology developers and transport fleet managers. Efforts from Intel Corporation are noteworthy in designing state-of-the-art vehicle fleet management systems that are geared towards optimizing fleet operations and maximizing the revenue opportunities. They are designed as end-to-end solutions for smart city transportation needs and are currently under operation in several countries [4] [5] [6].

However, such systems are overtly expensive and are not affordable by poor countries. Further, it was found that such devices lack the feature of monitoring the vehicle health as it is an essential parameter for the bus operators due to high investment in vehicle assets. This feature is also of vital importance to the drivers because in some cases the drivers themselves are vehicle managers and oversee their businesses operations. Nonetheless, this feature can be easily integrated in the design of such devices by taking aspiration from study conducted in [7]. Some innovative ideas of building low-cost models for vehicle tracking and providing decision information through various methods to passengers are proposed in [8][9][10] that are similar to the design objectives of this study, a solution feasible for poor countries.

3. Proposed System Model

The proposed solution is a device named as OBID (On board Intelligent Device) and was designed and continuously developed to leverage the accessibility of the public transportation by introducing technological intelligence in the buses currently running in Karachi, some of which are either old reconditioned buses or conventional run-of-the-mill buses, operating since early 1970s [11].

The essential task of this device is to gather, record and transfer real-time data to the back-office system. It is developed using common available components and open-source software for maintaining scalability while making it an ultra-low-cost, robust, accurate and extendable Internet of Vehicle data recording platform. This factor is very important as the platform shall be able to integrate in any of existing as well as the new model buses introduced in the public transit projects in such countries. The generated dataset is categorized in to three main branches as shown in figure 1. All data streams are sampled, timestamped, geotagged and then logged into internal SD card storage. Simultaneously, the data packet is also transmitted to an IoT cloud application as well to the MQTT broker, thereafter forwarded to database server for long-term storage.

4. Hardware and Software Design

The backbone of this IoT Gateway solution is the hardware and low-level firmware that performs the data collection and communication. It consists of a single board computer embedded system connected with digital and analog sensors, vehicle dashboard through CAN Bus and cellular network via HSDPA/WCDMA or GPRS/EDGE network. Raspberry Pi 3 is used as the main computing platform due to the availability of HDMI, Wi-Fi and Bluetooth interface and simplicity of use in a single board.

The developed test prototype is shown in figure 2 and the hardware components used in the proposed model are illustrated in figure 3. The software model is built upon the robust and reliable hardware architecture of OBIB and it handles all raw sensory data recordings and data transmissions to

IOP Conf. Series: Materials Science and Engineering **899** (2020) 012004 doi:10.1088/1757-899X/899/1/012004

cloud applications and data servers. It enables the framework of Internet of Vehicles in the public transportation systems – helping to transforming into a smart city.

4.1.1. System Script. After the system boots, the Rasperry Pi runs a series of scripts to perform its operational requirements, following the same procedure listed in [7] Software section. The main script, after initialization, proceeds to check the current status of the vehicle and subsequently shift into the required operating mode as shown in figure 4.

4.1.2. IoT Analytics and Visualization. The OBID software framework leverages on open source IoT connectivity protocols. The data pipeline journey relies on front-end and back-end interfaces. The back-end office system comprises of an MQTT based Eclipse Mosquitto broker. It is a machine-to-machine lightweight published-subscribe based protocol used for IoT messaging with low-power devices [12]. The fleet management and smart transportation model is based on three layers: Sensing Layer, Communications Layer and Services Layer. In the Sensing Layer the Gateway publishes metadata messages to the MQTT broker located at the bus station or central office, connecting the ground fleet with the back-office system. Similarly, several bus stations can be extended to connect with each other using publish-subscribe model at the Communications Layer, thus serving as a centrally connected data infrastructure. Finally, at the Service Layer the MQTT broker serves as the bridge for operation and

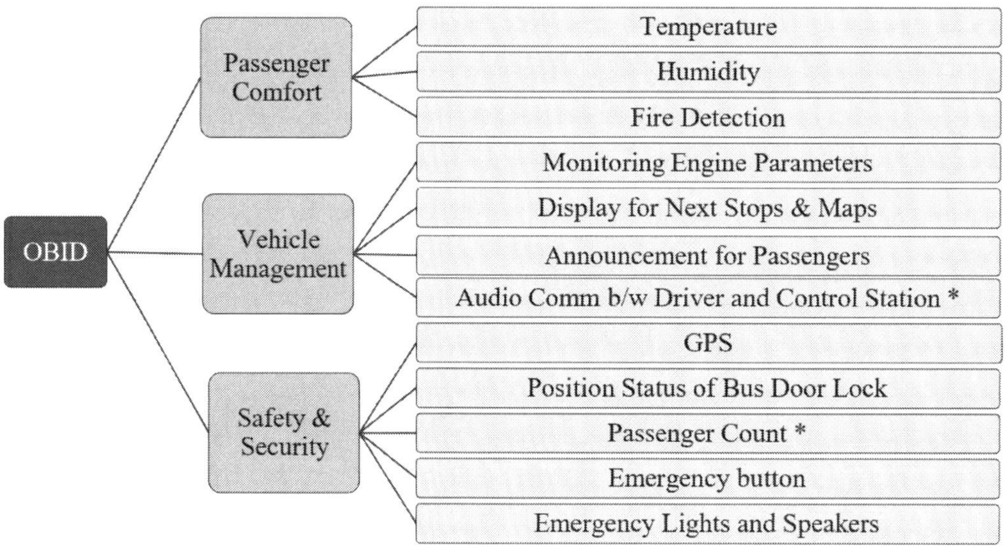

Figure 1. Set of features employed in OBIB. Note the task marked with '*' are still work in progress

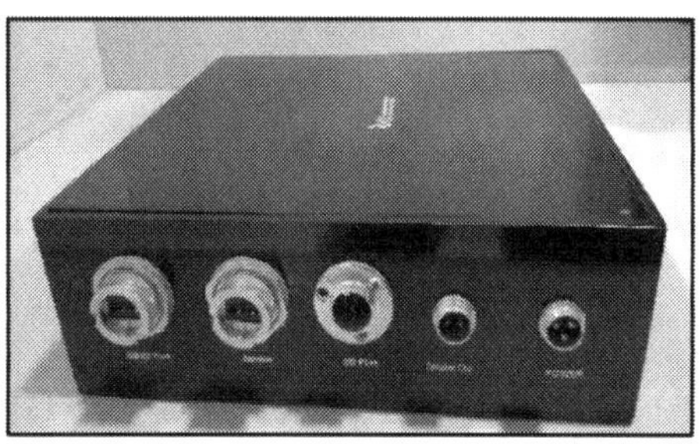

Figure 2. Prototype version of OBID. Showing hardware enclosed in a 5mm Plexiglass with weather-proof I/O, and power connectors

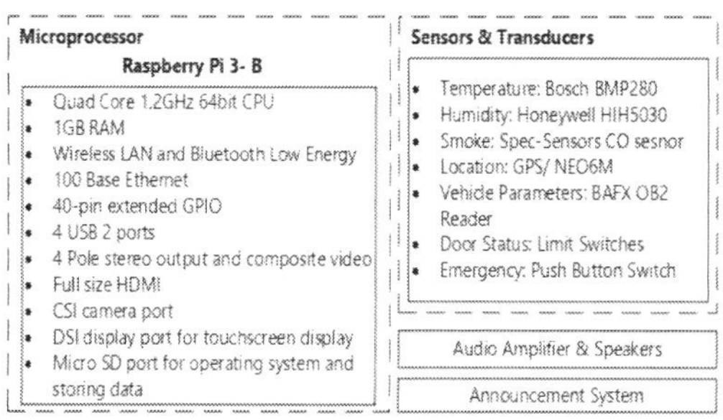

Figure 3. List of hardware components with specifications

continuous development of web services, mobile apps and business analytics. The data pipeline journey and communications model is illustrated in figure 5.

4.2. User Interface and Mobile App

The acquired data is sent to a third-party IoT platform, named Ubidots, which provides an internet service to display the data in a user-friendly dashboard and design mobile apps [13]. The collected data is sent via API structure provided by Ubidots. Based on a systematic requirement, three separate accounts were developed: Developers, Transport Operators and Passengers. Figure 6, 7 and 8 shows an instance of the passenger's mobile app, homepage and dashboard view for bus operators respectively.

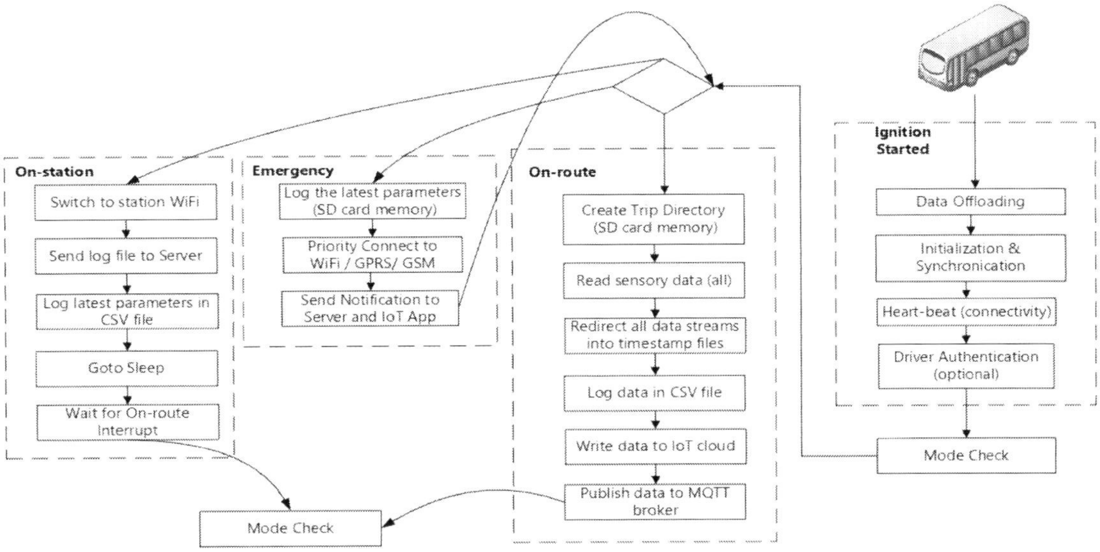

Figure 4. OBID main script procedure showing the process of multiple operating modes.

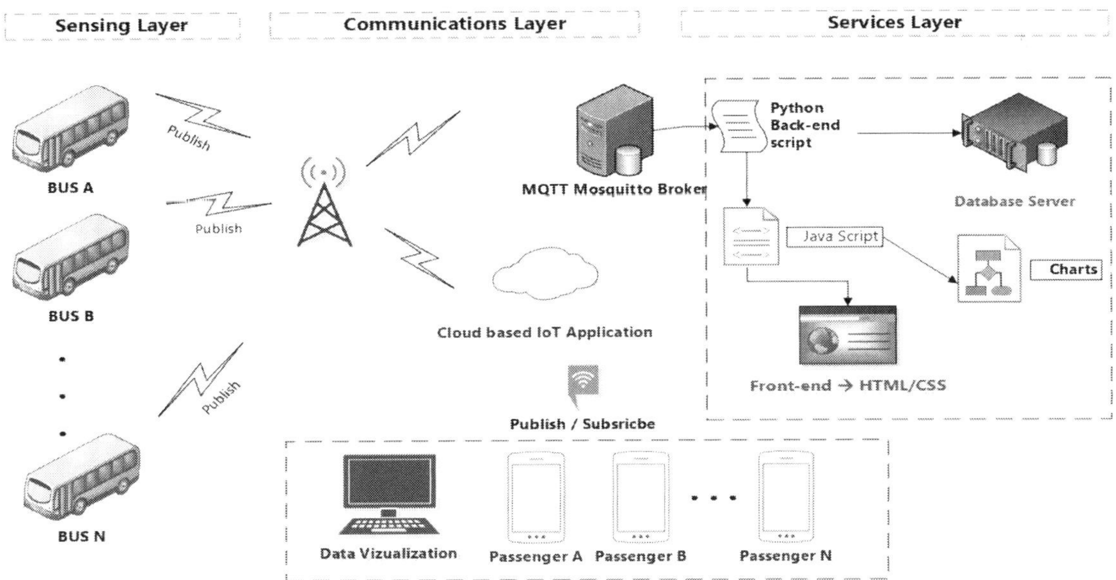

Figure 5. Developed Communication framework of OBID illustrating the data tranfer among the bus fleet, passengers and business owners.

4.3. Senor Mounting and Packaging

The peripherals and sensing hardware are placed on the bus ceiling. There can be several placement layouts with different number of sensors aiming to cover a particular cross section area of the bus. However, two sensors placed equal distance along the ceiling is found as optimal layout in the considered scenario. Despite the limitations in the coverage area, the combined average value of the two measurements lies in the acceptable tolerance rate, e.g., ±1°C in the case of ambient temperature. The CAD illustration in figure 9 shows the proposed sensor layout and device mounting in the bus, while figure 10 shows the illustrated component placement inside the OBID box.

5. Design Challenges

During the on-going development process and iterative field testing, the OBID has evolved into a robust platform of IoT Gateway in ITS applications. Each project section went through numerous iterations and we met some design challenges during the research, development and testing of this research project. This prototype was successfully instrumented with select vehicles along with a pilot test in Gulistan-e-Jauher, Karachi, Pakistan.

While the component placement inside the bus seems like a straightforward procedure, however, a great deal of efforts are required to securely place the box and other peripherals inside a public transport vehicle. The data and power cables have to be secure, hidden and of minimum length as possible to avoid signal degradation and power losses. With little to no information openly available about OBD protocols by vehicle manufacturer in the CAN bus, only a very minimal information can be captured by commercial OBD readers. To overcome this limitation, raw CAN data coming into the OBD port has to be recorded and interpreted. In some cases, the OBD port was found to be completely inaccessible by manufacturers due to warranty terms and conditions.

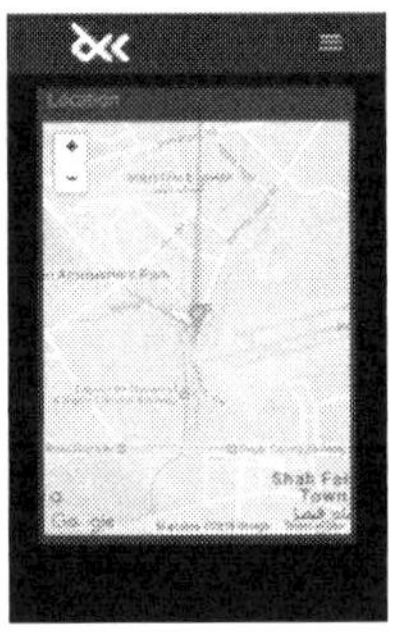

Figure 6. App for citizens.

Figure 7. Login webpage for bus operators.

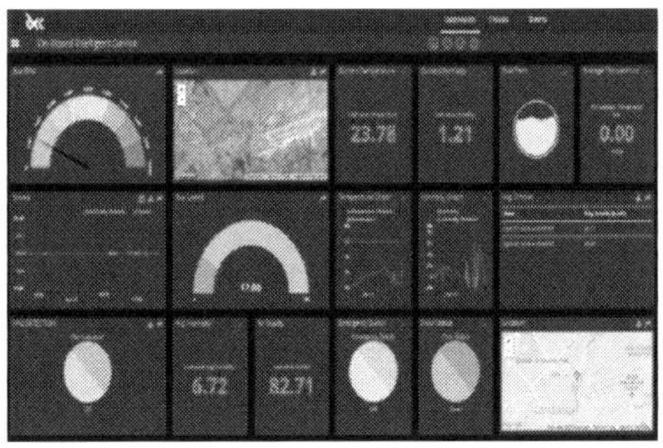

Figure 8. Dashboard screen showing field experiment recordings of a test vehicle.

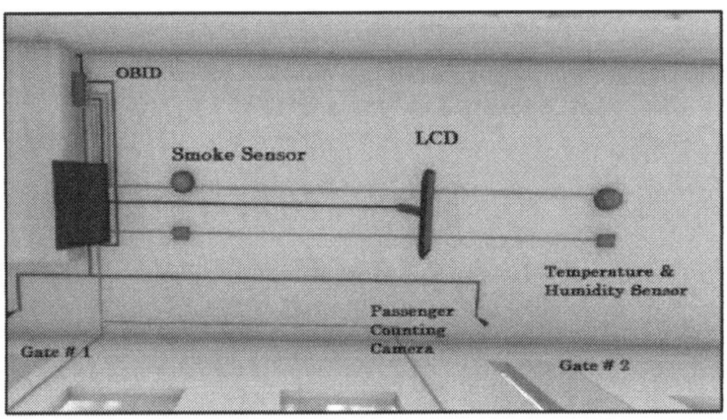

Figure 9. Proposed installation layout inside bus.

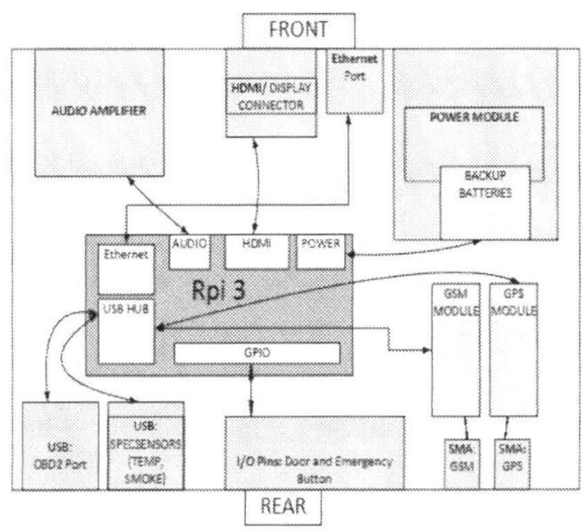

Figure 10. Component placement inside OBID box.

6. Future Work

The next generation of smart transportation solutions builds up on massive datasets generated from IoT gadgets installed in the transportation infrastructure. Recently, the use of AI in producing business insights are some of the areas that have attained considerable attention from business owners and technology developers Since, humans are not great drivers because we often multitask and take most actions based on emotional and social variables. Therefore, the prediction of driving patterns, in particular, is of great advantage to transport operators as the driver performance plays a key role in maintaining the health of the vehicles, which are one of the biggest assets of the business owners. It can help reducing the vehicle operating cost, contributing to less carbon emissions, and indirectly helping in achieving of a sustainable transportation model.

Moreover, the prediction of rush hours, analysis of average daily traffic volume, the speed of traffic, forecasting of road congestion and, a newly proposed concept of environmental parameters gathered from sensors mounted on vehicle's rooftop, can serves as a data bank for businesses and organizations. These results can help to make a positive impact on the city by helping the businesses and policy makers to identify key areas that need intervention. By providing centrally coordinated services this solution can make the cities safer, foster the gender equity and will boost private sector investment. Hence, the on-going future work includes the applications of computer vision and deep learning techniques on the data sets of transportation systems to yield useful insights that can benefit the social and commercial sector of the city.

7. Conclusion

ITS devices comprise of innovative hardware and software based solutions that aim to provide user services to passengers and transport managers to make safer and efficient decisions regarding their interaction with transportation infrastructure and services. The proposed solution demonstrates its feasibility for poor countries utilizing the existing infrastructure and requires minimal investment cost. It also does not require any additional civil works or procurement of new buses. The biggest advantage is that this device serves as a data collection tool for the government and transport authorities, acting as a primary step for mechanisms leading to reshaping the policymaking culture and service improvement. In addition, it also serves as a gateway to improved traffic flows by reducing commuting time delays, fuel wastage and air pollution.

8. References

[1] Tarapiah S, Shadi S, Nooraldin M, Sundos 2014 Offline public transportation management based on GPS/WiFi and OpenStreetMaps 2014 6th Int Conf on Computational Intelligence, Communication Systems and Networks (doi:10.1109/cicsyn.2014.46)

[2] Sharad S, Sivakumar P B, & Narayanan V A 2016 The smart bus for a smart city — A real-time implementation. 2016 IEEE Intl Conf on Advanced Networks and Telecommunications Systems (doi:10.1109/ants.2016.7947850)

[3] David M 2015 Road to safety: how the iot is driving transportation down a safer path Intel last accessed on 16th March 2020 https://blogs.intel.com/iot/2015/02/10/road-safety-iot-driving-transportation-safer-path/#gs.0jlx7v

[4] Transwiseway 2016 Data center architecture for internet of vehicles Intel Corporation last accessed on 28th June 2020 https://www.intel.com/content/www/us/en/big-data/xeon-e5-e7-transwiseway-paper.html

[5] Intel Corporation 2015 Building an intelligent transportation system with the internet of things (Iot) last accessed on 28th June 2020 https://www.scribd.com/document/358505142/iot-building-intelligent-transport-system-blueprint-pdf

[6] Transcode 2016 Transcode does smart transportation with Intel® Iot gateway Intel last accessed on 28th June 2020 https://www.intel.com/content/www/us/en/connected-transportation-logistics/smart-transportation-brief.html

[7] Lex F *et al* 2019 MIT autonomous vehicle technology study: large-scale naturalistic driving Study of Driver Behavior and Interaction With Automation *IEEE Access* **7** pp. 102021-102038 (Available: https://arxiv.org/abs/1711.06976)
[8] VR Channakeshava G and K G, 2015 Real time vehicle fleet management and security system *IEEE Recent Advances in Intelligent Computational Systems* (Trivandrum) pp. 417-421(doi: 10.1109/RAICS.2015.7488452)
[9] Ricardo S C and Alvaro P de la C 2018 Public transport vehicle tracking service for intermediate cities of developing countries, based on ITS architecture using Internet of Things (IoT), *21st Int Conf on Intelligent Transportation Systems Hawaii* pp. 2784-2789 (doi:10.1109/itsc.2018.8569906)
[10] Muhammad U F, Aamna. S and Abu B S 2017 GPS based public transport arrival time *prediction Int Conf on Frontiers of Information Technology Islamabad* pp. 76-81(DOI: 10.1109/FIT.2017.00021)
[11] Arif H and Mansoor R 2015 Karachi: The transport crisis *Karachi: Urban Resource Centre* pp. 13-15.
[12] "https://mosquitto.org/"
[13] "https://ubidots.com/"

Acknowledgments
The authors would like to thank Data Communication & Control (Pvt) Ltd Karachi for allowing to work on their project idea, especially Mr. Samir A. Hoodbhoy. This project was partially funded by School of Science & Engineering, Habib University. Especial thanks to vehicle owners who volunteered their vehicles during testing and development of this project.

IOP Conf. Series: Materials Science and Engineering **899** (2020) 012005 doi:10.1088/1757-899X/899/1/012005

Neutrino energy loss rates due to $^{66-71}Ni$ in stellar matter

Fakeha Farooq[1], Ramoona Shehzadi[2] and Jameel-Un Nabi[3]

[1,2]Department of Physics, University of the Punjab, 54590 Lahore, Pakistan
[3]GIK Institute of Engineering Sciences and Technology, Topi 23640, Khyber Pakhtunkhwa, Pakistan

Email address: fakehafarooq@gmail.com

Abstract. Rates for (anti-)neutrino energy loss on nickel isotopes, due to interactions involving weak decays ($\beta^{\pm}$-decay and lepton captures) are regarded as having fundamental importance during late evolutionary stages of massive stars. These rates substantially affect the leptonic ratio (Y_e) of stellar interior. For the densities less than 10^{11} g/cm^3, weak processes produce (anti-)neutrinos which cause reduction in the stellar core's entropy. In this paper, rates for neutrino and anti-neutrino energy loss on nickel neutron-rich isotopes ($^{66-71}$Ni) have been presented. Rates for energy loss have been determined by applying the deformed pn-QRPA model. The ranges for temperature and density, have been used to determine the rates, are from 0.01 to 30 (10^9 K) and 10^1 to 10^{11} (g/cm^3), respectively. Our computed rates for energy loss, at higher temperature regions, are enhanced in comparison with previously reported rates of Pruet and Fuller (PF).

1. Introduction

Stars are considered to be nucleosynthesis laboratories in the universe, which give out almost all the elements in the interstellar medium via the supernova explosions. Hydrodynamically stable massive stars (> 8 $M_\odot$) run through several burning phases and develop an iron core. At this stage the inward self-gravity force is balanced by the electron degeneracy pressure [1,2,3]. Further burning of silicon in shell exceeds the mass of core than the Chandrasekhar mass (~1.5 $M_\odot$), there the degeneracy pressure fails to overcome enormous gravity force and core is derived towards collapse forcibly [4, 5]. The gravitational core-collapse results in star's death through gigantic burst known as supernova (e.g. type II) explosions.

For the better understanding of the core-collapse supernova dynamics, the nuclear processes mediated by weak-interactions have considerable importance. The weak-processes; viz β-decays and lepton (electron and positron) captures cause reduction in the Y_e (fraction of lepton to baryon) and core entropy in the presupernova phase of massive stars [4, 6]. These processes are also responsible for an abundant production of (anti-)neutrinos up to ~ 10^{11} g/cm staller densities [7,8]. Energy budget of stars are decided by the (anti-)neutrinos which escape from stars and take away entropy and energy, resulting in cooling of the core. Therefore, it is valuable to calculate the rates for (anti-)neutrino energy loss for the Fe-peak nuclei. Throughout in this paper υEnL ($\bar{υ}$EnL) will be used for neutrino (anti-neutrino) energy loss.

The weak decay rates and the associated rates for ($\bar{υ}$)υEnL are ruled by the Gamow-Teller (GT) distributions strength [4]. In the presupernova and core-collapse phases, the GT transitions for Fe-peak nuclei are of central importance [9]. To precisely determine the GT transitions requires a reliable and microscopic theory. In this work, we are using the proton-neutron quasi-particle random phase approximation (pn-QRPA) [10] model for the calculations of ($\bar{υ}$)υEnL rates of nickel neutron-rich isotopes.

IOP Conf. Series: Materials Science and Engineering **899** (2020) 012005 doi:10.1088/1757-899X/899/1/012005

In the late evolutionary stages of massive stars, isotopes of nickel are produced abundantly and play a crucial role to change Y_e. As per studies of Ref. [11,12], the electron captures (EC) and decay rates of Ni isotopes have significant importance in astrophysical environment. Ref. [13] first reported the GT strength and EC rates of ^{56}Ni by using the pn-QRPA model. Later, the results were improved [14] and rates were calculated for heavier nickel isotopes (mass ranging 57-65). Ref. [15, 16] recently calculated the weak-interaction rates and GT transition properties on Ni isotopes. In this paper our focus is to calculate the rates for ($\bar{\upsilon}$)υEnL of nickel neutron-rich isotopes $^{66-71}$Ni in both leptons capture and lepton emission directions. The deformed pn-QRPA model has been used to compute the rates and our results have been compared with the earlier results of Pruet and Fuller (PF) [17].

In the next section, pn-QRPA has been described briefly. The detailed formalism can be found in Ref. [10]. Section 3 comprises discussion on our calculated rates and comparison of our results with other model calculations. Conclusion of our present work is given in Section 4.

2. Formalism

The Hamiltonian for the pn-QRPA model is

$$H^{QRPA} = h^{sp} + \upsilon^{pair} + \upsilon^{pp}_{GT} + \upsilon^{ph}_{GT} \qquad (1)$$

Nilsson model [18] was used to calculate the single-particle energies and wave functions. Pairing interactions υ^{pair} in nuclei were handled by using the BCS approximation. The residual interactions for proton-neutron pairs were included by υ^{pp}_{GT} (particle-particle) and υ^{ph}_{GT} (particle-hole) GT forces, respectively.

Among other necessary parameters for pn-QRPA model are the Nilsson potential parameters [18], the Nilsson oscillator constant $\Lambda = 41A^{-1/3}$ MeV, the pairing gaps $\Delta_n = \Delta_p = 12/\sqrt{A}$, the nuclear quadrupole deformation parameters [19], and the reaction Q-values [20].

For a transition from the kth parent to the mth daughter state, the ($\bar{\upsilon}$)υEnL rate of weak interactions is given by

$$\lambda^{N}_{km} = \left(\frac{\ln 2}{D}\right) f_{km}(T,\rho,E_f)\,[B(F)_{km} + (g_A/g_V)^2\, B(GT)_{km}],\quad N \equiv \upsilon,\bar{\upsilon} \qquad (2)$$

In equation (2) $B(F)_{km}$ and $B(\mathrm{GT})_{km}$ are reduced Fermi and GT transition probabilities, respectively. For current calculations, D=6143s (taken from [21]) and the value of ratio g_A/g_V is -1.2694 (taken from [22]). The f_{km} is the phase space integrals (natural units: h = c = m_e = 1), for electron and positron (upper and lower sign, respectively) emission

$$f_{km} = \int_{1}^{w_m} w(w^2-1)(w_m - w)^2 F(\pm Z, w)(1-\mathrm{G_m})\,\mathrm{d}w \qquad (3)$$

for electron (upper sign) and positron (lower sign) capture, the integral is

$$f_{km} = \int_{w_l}^{\infty} w(w^2-1)(w_m - w)^2 F(\pm Z, w)\,\mathrm{G_m}\mathrm{d}w \qquad (4)$$

In equations (3) and (4) w is total energy (kinetic plus rest) of electron or positron, w_l is the total threshold energy (kinetic plus rest) for electron or positron capture, and w_m is the total energy of β-decay. The $G_{\pm}$ are the Fermi-Dirac electron (positron) distribution functions and $F(Z,w)$ are Fermi functions.

The total ($\bar{\upsilon}$)υEnL rates are given by

$$\lambda^{N} = \sum_{km} P_k \lambda^{N}_{km}, \qquad N \equiv \upsilon,\bar{\upsilon} \qquad (5)$$

Here P_k is occupation probability of parent nucleus excited states which obeys the normal Boltzmann distribution and $\lambda^{N}_{km} = \lambda^{ec(pc)}_{km} + \lambda^{pe(ee)}_{km}$.

IOP Conf. Series: Materials Science and Engineering **899** (2020) 012005 doi:10.1088/1757-899X/899/1/012005

3. Results and discussion

In this unit, our estimated (ῡ)υEnL rates on $^{66-71}$Ni are presented. Tables 1 and 2 comprise the results of the deformed pn-QRPA computed (ῡ)υEnL rates for various densities region at different values of stellar temperature. The selected values for density ρ (g/cm^3) are given in the first column of the tables and the second column displays the chosen values of temperature T (10^9 K). For each isotope, two columns represent the rates for υEnL ($\lambda_{\upsilon}^{ec+pe}$) and ῡEnL ($\lambda_{\bar{\upsilon}}^{pc+ee}$), where ec (pc) and ee (pe) stands for electron (positron) capture and electron (positron) emission, respectively. The total (ῡ)υEnL rates take contributions from positron capture and electron emission (electron capture and positron emission).

Table 1. The pn-QRPA calculated rates due to $^{66-68}$Ni at certain densities and stellar temperatures. the terms $\lambda_{\upsilon}^{ec+pe}$ and $\lambda_{\bar{\upsilon}}^{pc+ee}$ are total (anti-)neutrino rates respectively in units of Mev/s.

ρ (g/cm^3)	T ($\times10^9$ K)	^{66}Ni $\lambda_{\upsilon}^{ec+pe}$	^{66}Ni $\lambda_{\bar{\upsilon}}^{pc+ee}$	^{67}Ni $\lambda_{\upsilon}^{ec+pe}$	^{67}Ni $\lambda_{\bar{\upsilon}}^{pc+ee}$	^{68}Ni $\lambda_{\upsilon}^{ec+pe}$	^{68}Ni $\lambda_{\bar{\upsilon}}^{pc+ee}$
	1.00	8.93E-57	6.92E-07	3.97E-49	9.53E-02	8.57E-64	2.40E-02
	5.00	7.29E-11	1.07E-02	1.00E-09	1.40E-01	1.68E-11	7.46E-02
1.00×10^2	10.00	8.89E-04	1.42E+00	1.36E-03	2.68E+00	6.10E-04	4.86E+00
	30.00	5.27E+02	3.32E+03	1.16E+03	6.11E+03	6.58E+02	6.78E+03
	1.00	1.03E-54	3.67E-07	4.56E-47	9.48E-02	9.84E-62	2.36E-02
	5.00	7.45E-11	1.05E-02	1.03E-09	1.39E-01	1.71E-11	7.36E-02
1.00×10^5	10.00	8.91E-04	1.42E+00	1.37E-03	2.68E+00	6.11E-04	4.86E+00
	30.00	5.28E+02	3.32E+03	1.16E+03	6.11E+03	6.58E+02	6.78E+03
	1.00	1.53E-44	1.68E-16	7.18E-37	1.67E-02	1.55E-51	2.85E-06
	5.00	1.17E-08	4.55E-04	1.63E-07	3.27E-02	2.72E-09	6.32E-03
1.00×10^8	10.00	5.02E-03	3.89E-01	7.71E-03	7.36E-01	3.45E-03	1.91E+00
	30.00	5.73E+02	3.07E+03	1.26E+03	5.65E+03	7.13E+02	6.27E+03
	1.00	5.71E+03	1.00E-100	1.02E+05	1.00E-100	1.44E+02	1.00E-100
	5.00	8.45E+03	1.99E-24	1.52E+05	6.89E-23	1.42E+03	3.57E-22
1.00×10^{11}	1 0.00	1.62E+05	1.25E-11	3.56E+05	2.67E-11	1.34E+05	2.30E-10
	30.00	1.41E+06	4.59E-01	3.26E+06	8.47E-01	1.82E+06	9.48E-01

Table 2. Same as table 1 but for $^{69-71}$Ni.

ρ (g/cm)	T ($\times10^9$ K)	^{69}Ni $\lambda_{\upsilon}^{ec+pe}$	^{69}Ni $\lambda_{\bar{\upsilon}}^{pc+ee}$	^{70}Ni $\lambda_{\upsilon}^{ec+pe}$	^{70}Ni $\lambda_{\bar{\upsilon}}^{pc+ee}$	^{71}Ni $\lambda_{\upsilon}^{ec+pe}$	^{71}Ni $\lambda_{\bar{\upsilon}}^{pc+ee}$
	1.00	6.75E-54	1.30E-01	1.49E-69	1.59E-01	1.48E-60	5.24E-01
	5.00	3.97E-11	9.10E-02	9.73E-13	2.46E-01	2.79E-11	9.12E-01
1.00×10^2	10.00	5.13E-04	1.37E+00	1.66E-04	8.41E+00	5.38E-04	8.38E+00
	30.00	7.21E+02	1.69E+03	5.47E+02	7.55E+03	6.56E+02	5.31E+03
	1.00	7.74E-52	1.30E-01	1.70E-67	1.57E-01	1.70E-58	5.20E-01
	5.00	4.06E-11	9.08E-02	9.95E-13	2.44E-01	2.84E-11	9.08E-01
1.00×10^5	10.00	5.14E-04	1.37E+00	1.67E-04	8.41E+00	5.41E-04	8.36E+00
	30.00	7.21E+02	1.69E+03	5.48E+02	7.55E+03	6.56E+02	5.32E+03
	1.00	1.22E-41	5.96E-02	2.68E-57	2.41E-03	2.67E-48	3.38E-02
	5.00	6.44E-09	4.42E-02	1.58E-10	3.75E-02	4.52E-09	3.55E-01
1.00×10^8	10.00	2.90E-03	7.55E-01	9.40E-04	4.40E+00	3.05E-03	4.20E+00
	30.00	7.82E+02	1.57E+03	5.94E+02	6.98E+03	7.11E+02	4.91E+03
	1.00	2.37E+04	2.01E-96	4.51E+03	1.00E-100	2.16E+03	7.13E-90
	5.00	2.56E+04	3.84E-21	5.05E+03	4.60E-21	1.08E+04	4.15E-20
1.00×10^{11}	10.00	1.24E+05	1.39E-10	4.71E+04	1.12E-09	1.16E+05	1.04E-09
	30.00	1.99E+06	2.43E-01	1.53E+06	1.09E+00	1.79E+06	7.78E-01

Tables 1 and 2 shows that when temperature ($\times 10^9$ K) changes from 1 to 5, the values of υEnL rates increase abruptly, in the domain of low- and medium-density. Above T = 5×10^9 K, a small increase occurs in the values of υEnL rates with the rise of core density. In general, $\bar{\upsilon}$EnL rates also increase as temperature increases at each density. For $\rho=1.00\times 10^{11}$ g/cm^3, the values of $\bar{\upsilon}$EnL rates are smaller in magnitude as compared to those in the regions of low- and medium-density, particularly at low temperature.

Next, we present the comparison of our estimated υEnL and $\bar{\upsilon}$EnL rates with the PF [17] rates for three selected Ni-isotopes ($^{66, 68, 71}$Ni for space consideration). Figure 1 illustrates the ratio of ($\bar{\upsilon}$)υEnL rates calculated by our model to the PF rates ($R_{\upsilon(\bar{\upsilon})}$). These ratios are plotted against stellar temperature values for four particular densities (10^2, 10^5, 10^8, 10^{11} g/cm^3).

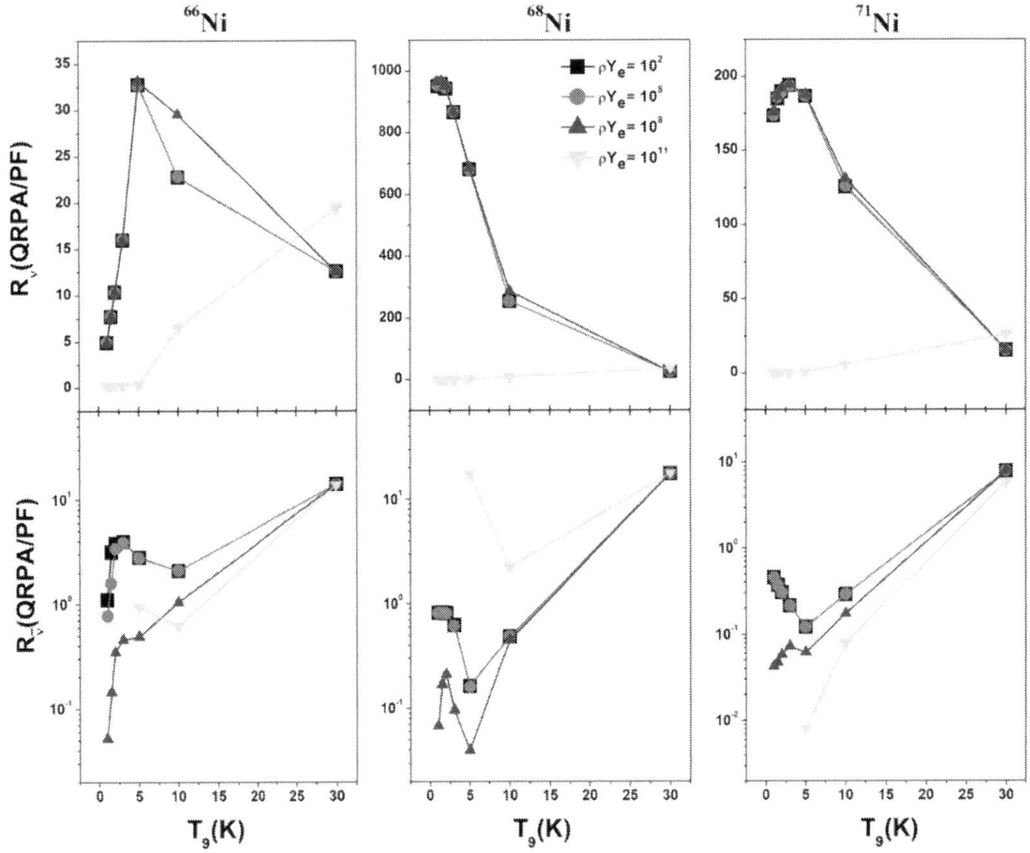

Figure 1. Ratios of calculation of ($\bar{\upsilon}$)υEnL rates due to $^{66, 68, 71}$Ni at selected temperatures and densities in stellar matter. R_υ (QRPA/PF) represents the ratios of our calculated υEnL rates to those of PF rates and $R_{\bar{\upsilon}}$ (QRPA/PF) for $\bar{\upsilon}$EnL rates. ρ is the baryon density in units of g/cm^3, Y_e is the leptonic ratio. T_9 ($T\times 10^9$ K) represents the temperature in stellar matter.

The first and second row of the figure 1 displays the ratios of neutrino (R_υ) and anti-neutrino ($R_{\bar{\upsilon}}$) rates, respectively. The graphical trend of R_υ shows that in the domain of low- and medium-densities, our computed neutrino υEnL rates are larger than those of PF rates. It can be noted from figure 1 that in case of anti-neutrino, at low temperature and for low- and medium-densities, the rates calculated by two models are in reasonable agreement. A large difference in the results of both pn-QRPA calculated rates and PF rates occur, at high density and low temperature (where values of rates are in itself very small). At high temperatures, both of the υEnL and $\bar{\upsilon}$EnL rates by using the deformed pn-QRPA are largely enhanced in comparison with PF rates.

IOP Conf. Series: Materials Science and Engineering **899** (2020) 012005 doi:10.1088/1757-899X/899/1/012005

4. Conclusion

We have determined the υEnL and $\bar{\upsilon}$EnL rates for Ni neutron-rich isotopes, over wide-ranging densities and temperatures. We also presented the comparison of our determined rates with the earlier Pruet and Fuller work. At higher stellar core temperatures (around 30×10^9 K) pn-QRPA rates are bigger by up to an order of magnitude than PF. We hope that our calculated ($\bar{\upsilon}$)υEnL rates may help in an improved comprehension of core-collapse simulations.

References

[1] Burbidge E M, Burbidge C R, Fowler W A and Hoyle F 1957 *Rev. Mod. Phys.* **29** 547
[2] Bethe H A 1990 *Rev. Mod. Phys.* **62** 801
[3] Rauscher T, Heger A, Hoffman R D and Woosley S E 2002 Nucleosynthesis in massive stars with improved nuclear and stellar physics *Astrophys. J.* **576** pp 323–348
[4] Bethe H A, Brown G E and Applegate J *et al* 1979 *Nucl. Phys. A.* **324** 487
[5] José J and Iliadis C 2011 *Rep. Prog. Phys.* **74**(9) 096901
[6] Langanke K and Martínez-Pinedo G 2003 *Rev. Mod. Phys.* **75** 819
[7] Heger A *et al* 2001 *Phys. Rev. Lett.* **86** 1678
[8] Janka H T *et al* 2007 *Phys. Rep.* **442** 38
[9] Fuller G M, Fowler W A and Newman M J *Astrophys. J. Suppl. Ser.* 1980 **42** 447 and 1982a **48** 279, *Astrophys. J.* 1982b **252** 715 and 1985 **293** 1
[10] Nabi J -U and Klapdor-Kleingrothaus H V 1999 *Eur. Phys. J.* A **5**, 337
[11] Aufderheide M B *et al* 1994 *Astrophys. J. Suppl.* **91** 389
[12] Heger A *et al* 2001b *Astrophys. J.* **560** 307
[13] Nabi J U and Rahman M U 2005 *Phys. Lett.* B **612** 190
[14] Nabi J U 2012 *Eur. Phys. J.* A **48** 84
[15] Nabi J U and Majid M 2017 *Int. J. Mod. Phys.* E **26** 1750005
[16] Cakmak S, Nabi J U and Babacan T 2018 *Nucl. Phys.* A **970** 86
[17] Pruet J and Fuller G M 2003 *Astrophys. J. Suppl. Ser.* **149** 189
[18] Nilsson S G 1955 *Mat. Fys. Medd. Dan. Vid. Selsk* **29** 16
[19] Möller P *et al* 1995 *At. Data Nucl. Data Tables* **59** 185
[20] Audi G *et al* 2017 *Chin. Phys.* C **41** 030001
[21] Hardy J C and Towner I C 2009 *Phys. Rev.* C **79** (5) 055502
[22] Nakamura K 2010 Particle Data Group *J. Phys. G: Nucl. and Part. Phys.* **37** (7A) 075021

Self-cleaning of glass surface to maximize the PV cell efficiency

Adnan Ayaz[1], Hussain Ahmad[2], Faraz Ahmad[3*], Ahmad Khan[4], S. M. hasnain Tarmazi[5], Rizwan M. Gul[6], Saim saher[7]

[1, 2, 4, 5, 6] Department of Mechanical Engineering, University of Engineering and Technology, Peshawar, 25000, KPK, Pakistan
[3]Department of Mechanical and Aerospace Engineering, Air University Islamabad, E-9, Islamabad Capital Territory 44000, Pakistan
[7]CECOS University of IT & Emerging Sciences, Hayatabad, Peshawar, KPK, Pakistan

*Corresponding Author Email Address: faraz.ahmad0460@gmail.com

Abstract. Photovoltaic (PV) modules are widely used for harnessing solar energy which ensure maximum output when their glass surface is clean. However, PV modules are open to dust, grime and other contaminations which get deposited on their surface causing reduction in transmittance and hence their efficiency reduces. It is therefore required to clean the glass surface of PV modules time to time either manually by labor or using some special arrangements such as automated systems. However, these techniques are either laborious or require extra energy. Therefore, another solution to offset such complications is to use chemical coatings which ensure self-cleaning of glass surface by increasing water contact angle. In the present study, two types of water repellent chemicals (such as trimethylchlorosilane and hexamethyldisilazane) have been used to coat the glass surface using dip coating technique. The performance of such coated glass slides has been investigated using some important characterization techniques, such as finding transmittance by spectrophotometer and measuring water contact angle using a high resolution camera. Moreover, the self-cleaning effect has been observed using a microbalance to measure dust on coated glass exposed to open atmosphere and compared with uncoated glass. The results revealed that these coatings have increased the water contact angle up to 149% which reduces friction between the glass surface and water droplets. Moreover, the friction reduction helps in mobility of water droplets which in turn can easily carry out dust along with them, thus improving the efficiency of PV module.

Keywords: Solar energy, photovoltaic, dust accumulation, chemical coating, water contact angle.

1. Introduction

Solar energy is a major source of renewable energy that is abundant in nature and free of cost, providing heat and electricity to many places across the globe. Being an important source of renewable energy, solar energy can be harnessed using a wide range of technologies such as solar architecture, photovoltaic, artificial photosynthesis, solar thermal energy, molten salt power plants and solar heating. Although solar thermal conversion provides a lot of benefits in term of heating in commercial applications, electricity generation from photovoltaic (PV) modules is attracting the focus of commercial, industrial and domestic users each passing day due to the higher grade of harvested energy. Its applications are becoming wide and wide starting from home applications to large industrial solar power plants.

1.1. Photovoltaic (PV) modules

Photovoltaic (PV) modules are used for harnessing solar energy and converting it to electrical energy. Due to its ability of harvesting both direct and diffuse sunlight, it can operate in both hazy and cloudy

IOP Conf. Series: Materials Science and Engineering **899** (2020) 012006 doi:10.1088/1757-899X/899/1/012006

conditions. PV modules are made up of solar cells which generate electricity working on the principle that light energy strikes the solar cell knocks electrons from atoms in semiconductor material thus producing electric current. Solar cells individually have low voltage and are combined into solar modules which are further combined to make solar panel and then a PV system. To meet the demand of specific power requirements, several panels are connected together to make a PV system as shown in Figure 1.

1.2. Photovoltaic phenomena

The PN-Junction solar cells use a semiconductor such as silicon which has four electrons in its valence shell. The conductance of pure silicon crystal lies in between that of pure conductors and pure insulators. Impurity atoms can be added to increase its conduction by a process called doping. Negative charges called electrons are introduced by doping with atoms containing more electrons than silicon. Being extra, these electrons are not tied up in chemical bonds and so are free to move around. Likewise, freely moving positive charges called holes are introduced by doping with atoms containing fewer electrons than silicon. Since these charges are mobile, they can carry an electric current. When an n-type and a p-type semiconductor come together, they form a PN-junction. Electrons can wander across the junction leaving behind static positive charges on the atoms in the solid. On the other side, they join with the holes and neutralize them. At the same time, holes can wander across the junction, leaving behind the static negative charges on the atoms in the solid. Near the PN-junction, the region where there are no charge carriers is called Depletion Zone. The separated positive and negative charges create an electric field across the depletion zone. When energy from the light is incident on the PN-junction, it dislodges an electron, creating a mobile electron and an extra mobile hole. The electric field makes the electron flow to n-type material and hole flow to p-type material as shown in Figure 2.

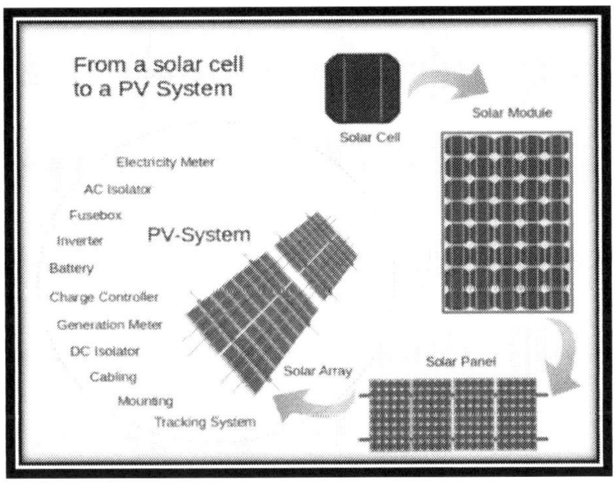

Figure 1. Combination of solar cells to make a PV system [1]

The resulting separation of positive and negative charges across the junction is called junction potential. Connecting the solar cell to an external circuit allows the electrons and holes to travel around and recombine returning the system to its starting condition. The charges driven around the circuit forms an electric current and does useful work such as lighting etc.

1.3. Factors affecting the performance of PV module

PV modules are installed in open atmosphere where its performance is influenced by outdoor conditions such as ambient temperature, irradiance level, dust accumulation, wind speed, and humidity etc. which have been addressed by a lot of researchers [3-7]. The performance of PV modules can be assessed by a lot of parameters in which efficiency is a prime factor. PV modules have generally low energy efficiencies; however, they ensure maximum achievable efficiency when its top surface is clean enough. The modules are generally mounted on ground and rooftop, deriving energy from solar radiations and are exposed to sunlight in open atmosphere thus eventually dust, grime and other contaminations get adhered to its surface which degrade its performance. Previous researchers proposed that dust is one of the most important factors that should be considered before installing a PV system.

1.4. Effect of dust on the performance of PV module

Dust accumulation is one of the major problems facing by PV modules to degrade its performance and life [8]. The reduction in efficiency of PV module due to dust accumulation depends on the size and mass of dust. It has been observed by Sarver et al. [9] that increase in mass of dust has a direct impact on efficiency degradation while size of dust has inverse impact on it. Darwish [10] investigated the effect of dust deposition on daily energy loss of PV module for one year. He examined that deposition of dust can cause 4.4% loss in mean of daily energy. He also proposed that this loss can be higher than 20% when there is no rain for a long period. Similarly, Said and Walwil [11] carried out some experiments to find out the effects of dust fouling on glass cover of PV module. They examined 35% reduction in spectral transmittance and 20% in overall transmittance. Furthermore, Kumar and Chaurasia [12] conducted research to investigate the effect of dust on PV module's efficiency in India and they examined the accumulation of dust can degrade the efficiency of PV module up to 40% in desert area.

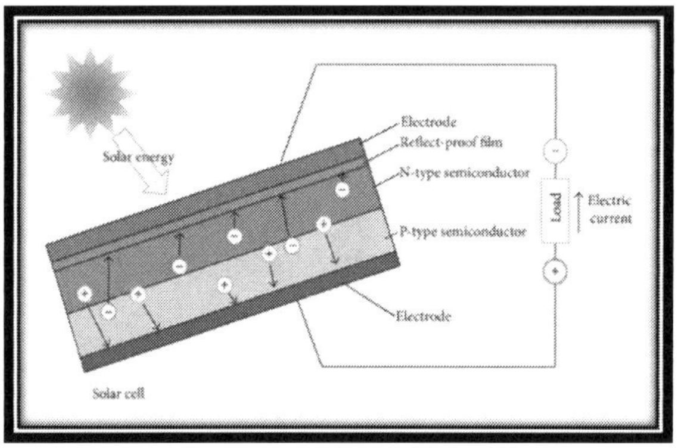

Figure 2. Photovoltaic phenomena [2]

Some researchers tried to generalize the effect of dust on performance of PV system and proposed some correlations between dust accumulation and efficiency of PV systems. For instance, Mekhilef et al. [3] carried out some experimental studies and found a correlation between the efficiency of PV module and thickness of the dust accumulated over its surface in a composite climate. They revealed that the accumulation of heavy layers of dust have a significant reduction of 10-20% in output of PV module while small amount of dust has a negligible effect on it. Similarly, Al-Hasan and Ghoneim [13] performed experimental and theoretical studies to correlate the efficiency of PV system with the quantity of dust accumulated. They measured and compared the I-V characteristics for both dusty and clean PV systems and proposed a linear correlation between the PV system's efficiency and dust accumulated on its surface. Moreover, they observed that the maximum output power and short circuit current decreases as the concentration of dust increases up to 1 g/m^2 but the rate of reduction decreases beyond this value. Furthermore, they concluded that there is no significant effect of dust accumulation on open circuit voltage of the PV system.

Different researchers attempted to prove that the effect of dust on PV system is different for different samples at different parts of the world. For example, Hussain et al. [14] carried out a detailed investigation of PV module subjected to different dust samples by analyzing its electrical performance such as output power, current and voltage. They observed that a minimum value of power was 3.88W when the PV module was accumulated by particles of rice husk. A similar type of research has been conducted by Kaldellis and Kokala [15] which concluded that the effect of dust on PV module performance varies from site to site. In this regard, Appels et al. [16] conducted research and proved that the problem of dust settlement in Belgium is not severe as compared to Middle East. Besides this, a very detailed review of dust soiling effects on solar panels has been done by Sayyah et al. [17]. They compared the effect of dust on solar panels for different parts of the world and concluded that the minimum power loss due to dust accumulation occurs in Cyprus while maximum occurs in Saudi Arabia.

Apart from investigating the effect of dust on PV module, some work has been dedicated to analyze its effect on transparent glass. For example, the effect of dust on transparent cover of solar collector has been discussed by Elminir et al. [18]. They observed that the transmittance of glass is greatly dependent on the deposition density of dust and tilt angle of solar collector. Moreover, they examined that the best tilt angle to reduce a maximum reduction of 17.4% in output power per month was 45° facing towards south.

1.5. Dust cleaning methods for solar panels

As discussed above, the accumulation of dust particles on the surface of PV module degrade its performance in terms of efficiency, output power and short circuit current etc. Therefore, proper functioning of PV module demands for some special arrangements that should be done to maintain its efficiency. In this regard, some techniques for cleaning PV modules has been summarized by He et al. [19] which include manual cleaning, natural cleaning, self-cleaning by coating with nano-films, electrostatic cleaning and mechanical dust cleaning methods etc.

Mostly, solar panels are cleaned manually using clean water and a little scrubbing with coarse cloth, covered sponge or soft brush to remove the most stubborn grime adhered to glass surface. However, this practice is very laborious and should be repeated time after time depending on the dust accumulation probability in that area where solar panel is installed. Similarly, dust can also be removed by natural process [20, 21] with the aid of heavy rains or strong winds. However, in the desert regions the winds will have an adverse effect in the form of increase in dust accumulation. Moreover, heavy rain will lead to non-uniform cleaning which leave permanent scratches on the surface of solar panel thus decreasing the power output. Another important method for solar panel cleaning is the use of automated system like Robots [22, 23]. The input to these systems can be given by making the use of a part of energy generated by PV system. Furthermore, the use of electrostatics force to remove sand from the surface of solar panels is currently in use in NASA's hover to Mars. A single-phase high voltage is applied to parallel wire electrodes embedded in the cover glass plate of a solar panel. More than 90% of the adhering sand is repelled from the surface with virtually zero power consumption [24, 25]. However, this method uses a lot of energy and can't be recommended for normal routines but rather they are used for special missions.

Apart from the above mentioned techniques, one of most promising technique is the use of self-cleaning nano-films which can be made by coating chemicals on the surface of glass of PV module. Chemicals possessing the property of self-cleaning have been under investigation for a long time because they can clean solar panel surfaces with minimum possible amount of water while keeping in view the importance of maintaining the transmittance of the glass.

1.6. Self-cleaning

The idea of self-cleaning of surface due to higher water contact angle was first discovered in lotus leaf known as lotus effect. Lotus leaf in order to wash out dust particles from its surface possesses the quality of super hydrophobicity which results in repellence of water droplets when falls on its surface. It has a water contact angle of more than 160° making its surface super hydrophobic [26]. Dirt particles are picked up by water droplets and hence results in self-cleaning. The image of water droplet on lotus leaf is shown in Figure 3. Inspired from the phenomenon of natural self-cleaning, many researchers have tried and developed artificial self-cleaning surfaces by applying certain chemical coatings.

1.7. Use of chemical coatings for self-cleaning

Chemicals are found in three states i.e. solid, liquid and gas and a variety of chemicals are available today. However, more research studies are being carried out and new chemicals are being introduced day by day because they play a vital role in modern engineering studies and engineers can't deny the importance of chemicals in modern applications. Various industries use chemicals for various applications such as foods, Beverages, paints and varnishes etc.

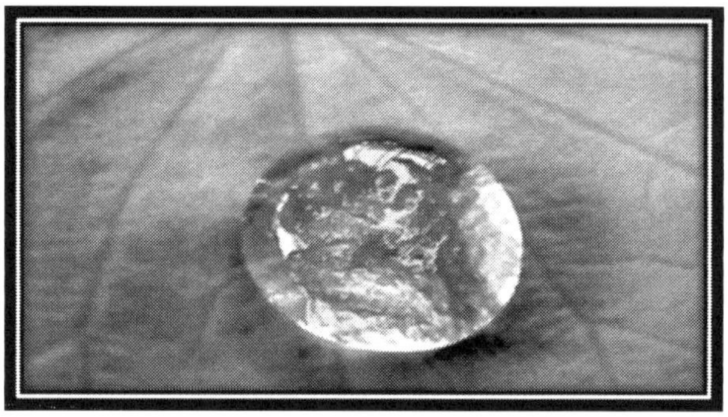

Figure 3. Lotus effect

Apart from daily life items, chemicals are used for different purposes in different industries. For example, in glass industries chemicals are coated on the glass surface to change its properties like transmittance, reflection and to decrease the roughness of glass surface and sometimes used to make glass surface attractive or repellent for water. Depending on the later applications, chemicals can be classified into following categories:

1.7.1. Hydrophobic chemicals

Hydrophobic means "water-hating". Sometimes water spreads evenly when it hits a surface; sometimes it beads into tiny droplets. Materials that naturally repel water, causing droplets to form, are known as hydrophobic. If the droplet forms a sphere that barely touches the surface like drops of water on a hot griddle, the contact angle is more than 90 degrees and the surface is hydrophobic, or water-fearing. Chemicals which are used to make any surface hydrophobic are called hydrophobic chemicals. Special techniques are applied while coating chemicals on substrate. When hydrophobic chemicals are coated on a surface and water comes in contact with that surface, droplets of water becomes spherical and tend to slip on the surface as shown in Figure 4. This property of surface refers to hydrophobicity and has been used by a lot of researchers to increase the efficiency of PV system [27-33].

Figure 4. Behavior of water droplet on hydrophobic surface

1.7.2 Hydrophilic chemicals:

Hydrophilic means "water-loving". Materials with a special affinity for water — those it spreads across, maximizing contact are known as hydrophilic. Hydrophilic and hydrophobic materials are defined by the geometry of water on a flat surface specifically the angle between droplet's edge and the surface underneath it which is called the contact angle. If the droplet spreads, wetting a large area of the surface, then the contact angle is less than 90 degrees and that surface is considered as hydrophilic surface. Chemicals coated to make any surface hydrophilic are called hydrophilic chemicals. When hydrophilic chemicals are coated over a surface and water comes in contact with that surface, it spreads over surface and tends to increase wettability as shown in Figure 5. This property of surface refers to hydrophilicity which has been used in a lot of studies to improve the performance of solar panels by preventing it from dust accumulation [34-40].

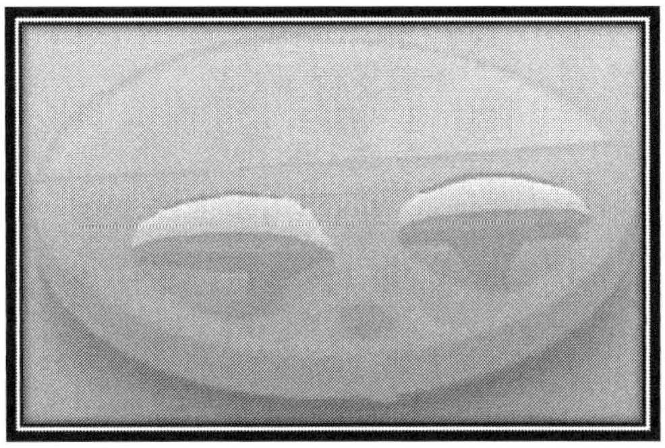

Figure 5. Behavior of water droplet on hydrophobic surface

2. Materials and methods

2.1. Selection of chemicals and coatings

There are different kinds of chemicals available for different applications. Hydrophobic and hydrophilic chemicals are being used by different industries for various sorts of items requiring certain hydrophobic or hydrophilic characteristics. Glass industry is one of the major users of these chemicals. Different types of hydrophobic and hydrophilic chemicals are available in the market today. In the present study, Trimethylchlorosilane (TMCS) and Hexamethyldisilazane (HMDS) were used to be coated on the glass surface whose thermo physical properties are given in Table 1.

Table 1. Thermo physical properties of chemicals

Properties	TMCS	HMDS
Form	Liquid	Liquid
Odor	Pungent	Ammonia like
Freezing point	-58 °C	-78 °C
Boiling point	56 °C	125 °C
Flash point	-27 °C	25 °C
Density at 20 °C	0.856 g/cm^3	0.774 g/cm^3
Molecular weight	108.64 g/mol	161.39 g/mol

TMCS is also known as Chlorotrimethylsilane which is a colorless fuming liquid with a pungent odor. Its molecular formula is $(CH_3)_3SiCl$ which reacts vigorously and exothermically with water. It is highly flammable and produces fumes in air. The vapor and liquid may cause burns. Hexamethyldisilazane is found in liquid form whose molecular formula is $C_6H_{19}NSi_2$. It is flammable and is miscible with acetone, ethyl ether and benzene. It is stable under normal temperatures and pressures but may decompose on exposure to moist air or water. It is very harmful if swallowed, inhaled or absorbed through skin.

A lot of techniques are available for chemical coatings such as chemical vapor deposition, physical vapor deposition, dip coating, spin coating and spraying etc. Although every technique has its own merits but dip coating was used in the present study because of its simple approach. Two separate dry and clean beakers were taken and labeled before making solution. Fairly cleaned glass specimens were labeled and made ready for coating. Both the chemicals (TMCS and HMDS) were used as solute and solution was prepared in the separate beakers by constant stirring for fifteen minutes with the help of a stirrer as shown in Figure 6. Benzene was used as solvent for TMCS and the ratio of solute to solvent was 1:5. Similarly, ethyl alcohol was used as solvent for HMDS in the ratio of 1:13. After making solutions, different glass samples were dipped with uniform speed and kept in the solution for thirty minutes. After thirty minutes glass samples were taken out of the beakers and were allowed to drain off excess amount of solution. Samples were exposed to dry environment and placed for some time to dry. Coating was carried out in this way and samples were made available for further testing.

2.2. Characterization techniques

The aim of characterization techniques is to forecast the behavior of solar panel when it is operational and exposed to dust, grime and other contaminants. Following techniques were selected for testing the effects of coating.

Figure 6. Preparation of solution for coating

2.2.1. Measurement of water contact angle (WCA)

Contact angle is an angle that a liquid creates with the solid surface when it is deposited on it. The contact angle depends upon both the properties of the solid as well as the liquid and the attractive and repulsive forces between liquid and solid. These interactions between solid and liquid are referred as the phenomena of cohesion and adhesion forces. Figure 7 shows the contact angle formed by a water droplet with the surface of a solid 2-D.

Contact angle defines the wettability of the surface. For example, the smaller contact angle means that cohesive forces are weaker than adhesive forces and molecules of liquid tend to interact more with solid molecules than liquid molecules and result in complete wettability of the surface. Generally such surfaces are known as hydrophilic surfaces with contact angle less than 90°. In other words these surfaces are the ones which attract water. There are some surfaces which exhibit extremely low contact angle approaching 0° are called super hydrophilic surfaces. Moreover, the larger contact angle means that cohesive forces are stronger than adhesive forces and the molecules of liquid tend to interact more with each other than with the solid molecules and result in poor wettability of the surface. Generally such surfaces are known as hydrophobic surfaces with contact angle greater than 90°. In other words these surfaces are the ones which repel water. There are some surfaces which exhibit extremely high contact angle approaching 180° are called super hydrophobic surfaces.

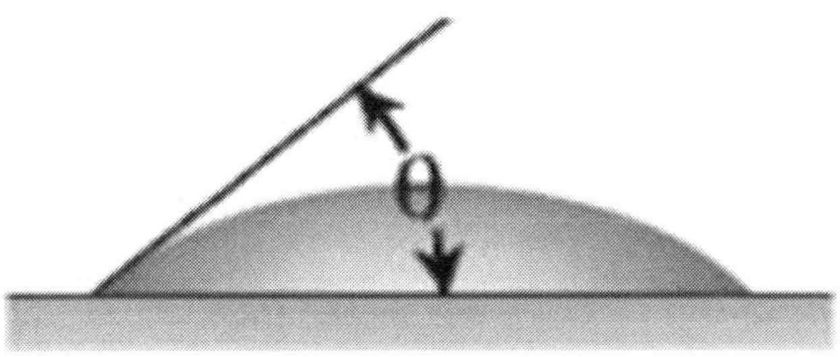

Figure 7. Measurement of WCA

The apparatus used for measuring WCA is Goniometer which consists of high resolution camera and a stand with adjustable height on which glass slide is placed. Droplet of water is poured on the surface and high resolution pictures are taken and analyzed in software. Since goniometer was not available in Pakistan so, a new apparatus was formulated for WCA measurements which is similar to goniometer as shown in Figure 8. High quality pictures were taken and converted into grey scales which are then analyzed by the online available software "image j".

Figure 8. Apparatus for measuring WCA

2.2.2. UV-Vis-NIR Spectroscopy

Ultraviolet visible near-infra red (UV-Vis-NIR) spectroscopy is a useful technique to characterize the absorption, transmission and reflection properties of a variety of technologically important materials, such as pigments, coatings, windows, and filters. This more qualitative application usually requires recording of at least a portion of the spectrum for characterization of the optical or electronic properties of materials. After chemical coating of glass slides by HMDS and TMCS, spectrometry of glass slides was conducted to found its transmittance. Two samples each of TMCS and HMDS and one uncoated glass slides were tested in Physics department at Material Research Lab. University of Peshawar, KPK, Pakistan.

2.2.3. Dust measurement

After coating samples with TMCS and HMDS, glass slides were kept for drying at room temperature. The weight of each glass slide was measured by microbalance (KERN, ALS 220-4) as shown in Figure 9. The sensitivity of microbalance was in range of 0.0001 g to 220g. Small plastic bags were weighted before use to carry glass slides. The glass slides were kept in open atmosphere under a controlled environment so that self-cleaning can only occurs by means of natural wind. Total of 30 glass slides (10 sets, each containing 1 TMCS, 1 HMDS and 1 uncoated glass sample) were placed for dust measurement test. Glass samples were numbered as shown in Figure 10 where any number without alphabet represents uncoated sample, alphabet A represents sample coated with TMCS and B represents sample coated with HMDS.

Figure 9. Microbalance (KERN, ALS 220-4)

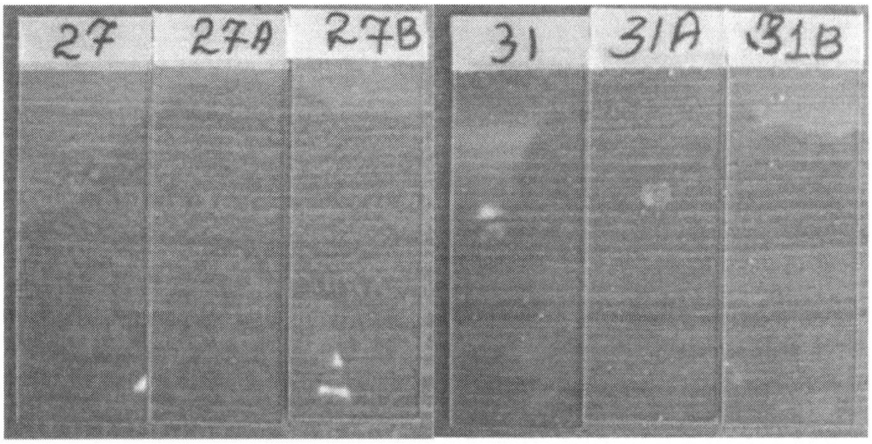

Figure 10. Numbering of glass slides

After 15 days, 5 sets of samples were selected to measure the total weight of the glass slide. Then the weight of the dust was calculated by subtracting the initial weight (weight of glass slide + weight of plastic packets) from the final weight.

2.2.4. Self-cleaning with the aid of water

Although the main purpose of such coatings is to achieve self-cleaning of dust particles in the presence of natural wind but sometimes it may not work well in some areas where wind velocity is very slow like plain areas of Khyber Pakhtunkhwa, Pakistan. Therefore, to evaluate and proof the possibility of self-cleaning, another test can be performed which is to analyze the self-cleaning phenomenon under the action of water. The weight of dust sprinkled on each sample was measured by means of microbalance which was around 0.4506 grams and the particle size was 74 microns which was measured with the help of a sieve. The samples were placed with 20° angle of tilt so that water droplets can easily roll over its surface under the action of gravity as shown in Figure 11. Then same amount of water (5 drops per sample=1ml) was poured by means of a dropper and self-cleaning was observed with naked eye.

Figure 11. Measuring tilt angle of glass sample

3. Results and discussion

3.1. Comparison of water contact angle

Water contact angle predicts the behavior of water droplet on a solid surface. Total of 30 glass slides (10 sets, each containing 1 TMCS, 1 HMDS and 1 uncoated glass sample) were taken and their water contact angles were measure and compared. Images taken for 1 set of glass slides are shown in Figure 12. For each category, an average value of 10 sets has been taken and compared with each other as shown in Table 2. It is clearly shown that TMCS sample has achieved the maximum average WCA followed by HMDS and uncoated sample. The high WCA of 68.96° is because of high non-polar nature of TMCS as compared to HMDS. The percent increase in WCA for TMCS was 149% while in case of HMDS, it was 48%. Although the WCA for both coatings was less than 90° showing that the surfaces are still hydrophilic but the increase in WCA means that it has increased the hydrophobicity of the glass surface. On the basis of such results, it can be concluded that TMCS can be used to enhance the hydrophobic behavior of any solid surface especially for those surfaces having the WCA greater than 90°. Moreover, the maximum WCA in case of TMCS coating witnessed that it has better hydrophobic behavior than HMDS and uncoated sample.

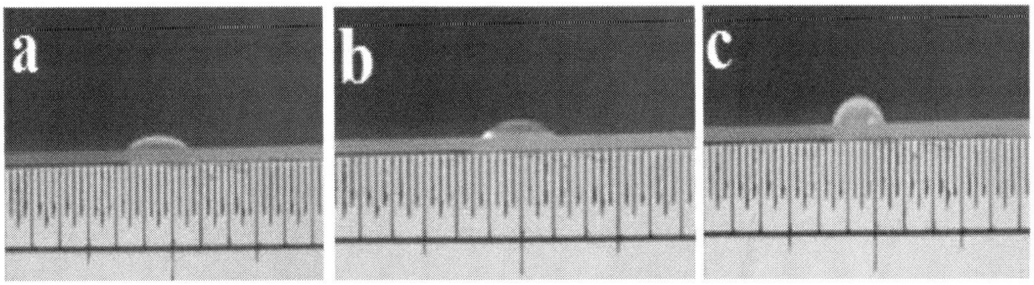

Figure 12. Water contact angles of (a) uncoated sample, (b) HMDS sample, (c) TMCS sample

Table 2. Comparison of water contact angle (degrees)

Sample No.	WCA for Uncoated sample	WCA for HMDS	WCA for TMCS
1	30.75	38.26	65.46
2	25.88	38.50	65.96
3	32.87	38.75	66.12
4	23.98	40.78	67.20
5	29.99	40.90	69.19
6	28.32	41.41	69.51
7	26.52	41.97	70.53
8	27.02	42.34	71.03
9	24.22	42.50	71.10
10	26.77	43.34	73.51
Mean	27.63	40.87	68.96

3.2. Spectrometry

The chemical coating over a glass surface may affect its transmittance characteristics. In this regard, UV-Vis-NIR spectrometry was utilized to investigate the effect of TMCS and HMDS coating on the transmittance of glass slides in electromagnetic spectrum having wavelength range of 250-900nm which is the working range of PV modules. Figure 13 shows the variation of percent transmittance and solar intensity with wavelength for uncoated, HMDS and TMCS coated samples. It can be noticed that percent transmittance increases when the wavelength increases from 250nm to 320nm. Moreover, the percent transmittance becomes almost constant when wavelength further increases from 320nm. It is also obvious that maximum transmittance has been observed for uncoated glass slides i.e. up to 90%.

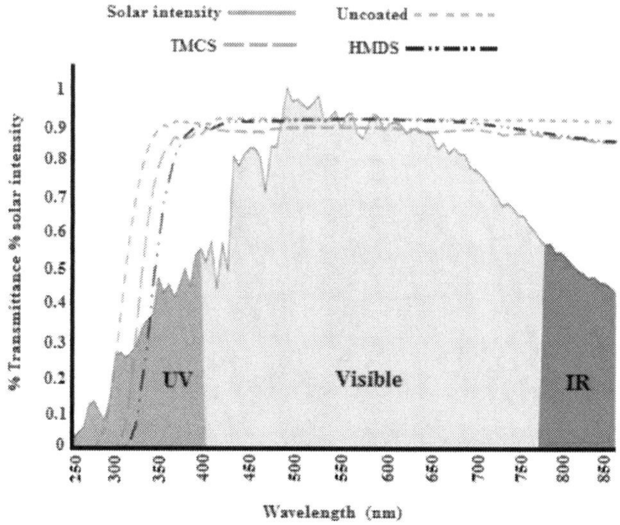

Figure 13. Variation of % transmittance and solar intensity with wavelength

Furthermore, it has been observed that the transmittance of coated glass slides starts later than uncoated glass slides in the UV region which improves the efficiency of PV module. It is due to the fact that UV light heats up solar cells, which causes reduction in the performance of PV module. In the visible region, the transmittance of coated samples is relatively less than that of uncoated sample. This reduction in transmittance can be compensated by the increase in WCA of coated samples.

3.3. Dust measurement

The main aim of this study was self-cleaning of dust with the aid of natural wind but it was examined that there was no significant effect of self-cleaning through wind in open atmosphere. To investigate this effect, 10 sets of glass slides were divided into two groups and both were placed at the rooftop of University of Engineering and Technology (UET) Peshawar, KPK, Pakistan. For the first 5 sets of glass slides, the amount of dust accumulated was measured after 15 days and compared with each other as given in Table 3. Although there was no significant difference between the dust accumulated on coated and uncoated samples; however, the maximum dust was accumulated over HMDS samples followed by TMCS and uncoated samples. Moreover, it was observed that the coated sample accumulates more dust over its surface as compared to uncoated sample after 15 days.

Table 3. **Comparison of dust accumulation after 15 days**

Sample No.	**Dust on Uncoated sample (g)**	**Dust on HMDS sample (g)**	**Dust on TMCS sample (g)**
1	0.4625	0.4587	0.4652
2	0.4271	0.4787	0.4486
3	0.4440	0.4700	0.4680
4	0.4476	0.4350	0.4502
5	0.4582	0.4834	0.4498
Mean	0.4479	0.4652	0.4564

Similarly, for next five sets of glass slides, the amount of dust accumulated was measured after 30 days and compared with each other as given in Table 4. It is clear that the maximum dust was accumulated on uncoated sample followed by HMDS and TMCS after 30 days which means that coatings helped to prevent some amount of dust accumulation which was not so after 15 days reading.

Table 4. Comparison of dust accumulation after 30 days

Sample No.	Dust on Uncoated sample (g)	Dust on HMDS sample (g)	Dust on TMCS sample (g)
1	0.5555	0.6315	0.6745
2	0.5979	0.6941	0.5831
3	0.6666	0.7050	0.6318
4	0.7161	0.5655	0.6302
5	0.6873	0.6042	0.6735
Mean	0.6445	0.6400	0.6386

Although self-cleaning through wind is not significant in the present study but the comparison of Table 3 and Table 4 shows that the effect of self-cleaning increases as the time span increases. Therefore, it can be concluded that TMCS and HMDS coating take some time to show their effects of self-cleaning through natural wind.

3.4. Self-cleaning with water

Although the main purpose of this study was to achieve self-cleaning of dust particles in the presence of natural wind but the results obtained was not significant. Therefore, another procedure was performed to examine the effect of coating on self-cleaning with the aid of water. In this regard, specific amount of water was sprinkled on each glass slides and the then water droplets were poured on them and self-cleaning was observed as shown in Figure 14.

Self-cleaning with the aid of water clearly shows that TMCS sample ensures maximum cleaning of dust as compared to HMDS and Uncoated glass. Moreover, no trail of water was left in the case of TMCS while mud is formed on uncoated and HMDS samples. Therefore, it can be concluded that TMCS is a promising candidate for self-cleaning under the action of water. This is probably due to high water contact angle for TMCS sample which reduce the friction for same amount of water droplet as on HMDS and uncoated surface. This reduction in friction will result in ease of mobility of water droplet which in turn can easily carry out dust along with it.

4. Conclusions

In the present study, an attempt has been made to prevent the surface of PV module from dust accumulation by use of chemical coatings. The chemicals used in this study were trimethylchlorosilane (TMCS) and heaxmethyldisilazane (HMDS) and tests were performed using glass slides for coating. The performance of glass slides coated with TMCS and HMDS was compared with each other and with uncoated glass slides. For this comparison, some important characterization techniques were used such as measurement of water contact angle (WCA), spectrometry, dust measurement, and self-cleaning with water. Some important conclusions that have been drawn from this study are:

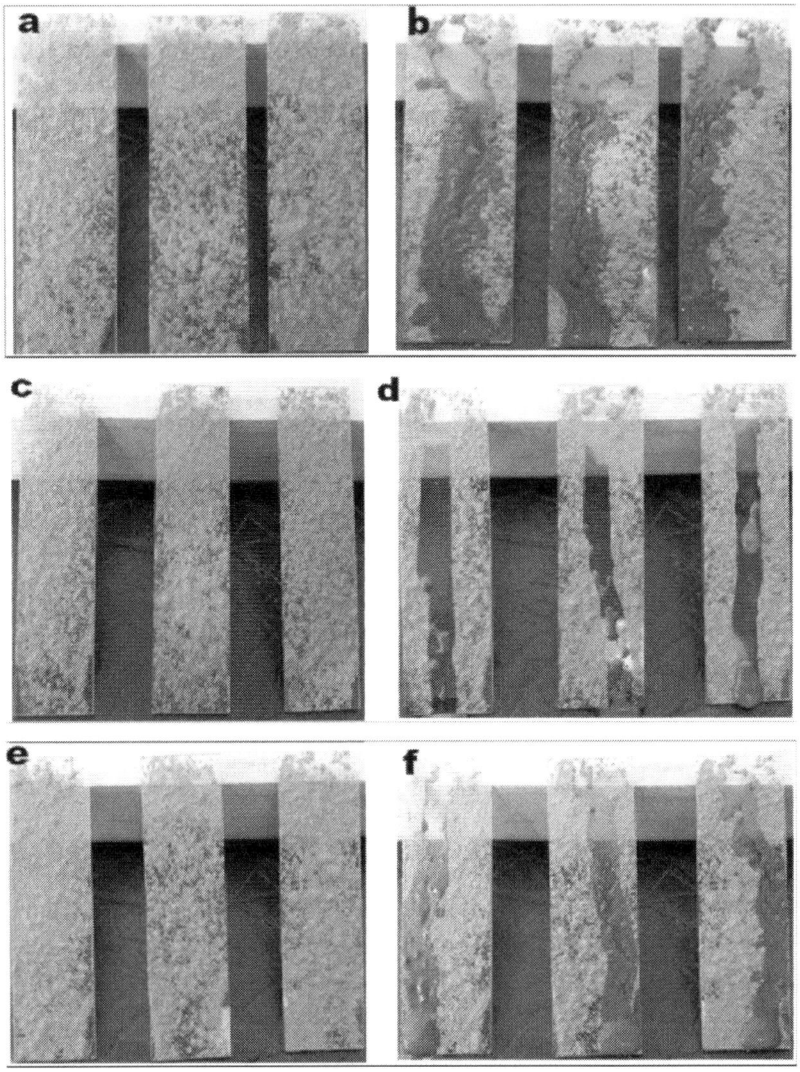

Figure 14. Comparison of self-cleaning effect for (a, b) uncoated sample, (c, d) TMCS sample, (e, f) HMDS sample

- The maximum WCA has been achieved by samples coated with TMCS followed by HMDS and uncoated sample. The high WCA of 68.96° is because of high non-polar nature of TMCS as compared to HMDS. Moreover, the percent increase in WCA for TMCS was 149% while in case of HMDS, it was 48%
- Although the WCA for both coatings (TMCS and HMDS) was less than 90° showing that the surfaces are still hydrophilic but the increase in WCA means that it has improved the hydrophobicity of the glass surface. On the basis of such results, it can be concluded that TMCS can be used to enhance the hydrophobic behavior of any solid surface especially for those surfaces having the WCA greater than 90°
- It has been observed that the transmittance of coated glass slides starts later than uncoated glass slides in the UV region which improves the efficiency of PV module. It is due to the fact that UV light heats up solar cells, which causes reduction in the performance of PV module
- The percent transmittance of TMCS and HMDS sample is slightly less than uncoated sample which is due to coating layer that has been deposited over glass slides during coating. However,

this reduction in transmittance is not so damaging as compared to the expected reduction in transmittance of solar panel with the passage of time when exposed to dust in atmosphere without coating

- Although self-cleaning through wind is not significant in the present study but the comparison of data for 15 days and 30 days shows that the effect of self-cleaning increases as the time span increases. Therefore, it can be concluded that TMCS and HMDS coating take some time to show their effects of self-cleaning through natural wind
- The increase in WCA causes reduction in friction that result in ease of mobility of water droplet which in turn can easily carry out dust along with it. Self-cleaning with the aid of water clearly shows that TMCS sample ensures maximum cleaning of dust as compared to HMDS and Uncoated glass. Moreover, no trail of water was left in the case of TMCS while mud is formed on uncoated and HMDS samples

References

[1] Umoh E A and Lugga A A 2016 Constraints to sustainable utilization of photovoltaic energy systems in built environment of kaura-namoda *NSE/UNESCO/FAEO Africa Engineering Conference on Energy* Egypt **2016**

[2] Zhang H, Van Gerven T, Baeyens J and Degrève J 2014 Photovoltaics: reviewing the European feed-in-tariffs and changing PV efficiencies and costs *The Scientific World Journal* **2014**

[3] Mekhilef S, Saidur R and Kamalisarvestani M 2012 Effect of dust, humidity and air velocity on efficiency of photovoltaic cells *Renewable and sustainable energy reviews* **16** 2920-5

[4] Meral M E and Dincer F 2011 A review of the factors affecting operation and efficiency of photovoltaic based electricity generation systems *Renewable and Sustainable Energy Reviews* **15** 2176-84

[5] Park N, Oh W and Kim D 2013 Effect of temperature and humidity on the degradation rate of multicrystalline silicon photovoltaic module *International Journal of Photoenergy* **2013**

[6] Ramli M A, Prasetyono E, Wicaksana R W, Windarko N A, Sedraoui K and Al-Turki Y A 2016 On the investigation of photovoltaic output power reduction due to dust accumulation and weather conditions *Renewable Energy* **99** 836-44

[7] Ye J Y, Reindl T, Aberle A G and Walsh T M 2014 Performance degradation of various PV module technologies in tropical Singapore *IEEE Journal of Photovoltaics* **4** 1288-94

[8] Rajput D S and Sudhakar K 2013 Effect of dust on the performance of solar PV panel *International Journal of ChemTech Research* **5** 1083-6

[9] Sarver T, Al-Qaraghuli A and Kazmerski L L 2013 A comprehensive review of the impact of dust on the use of solar energy: History, investigations, results, literature, and mitigation approaches *Renewable and sustainable energy Reviews* **22** 698-733

[10] Darwish Z A, Kazem H A, Sopian K, Alghoul M and Chaichan M T 2013 Impact of some environmental variables with dust on solar photovoltaic (PV) performance: review and research status *International J of Energy and Environment* **7** 152-9

[11] Said S A and Walwil H M 2014 Fundamental studies on dust fouling effects on PV module performance *Solar Energy* **107** 328-37

[12] Kumar S and Chaurasia P 2014 Experimental study on the effect of dust deposition on solar photovoltaic panel in Jaipur (Rajasthan) *Int. J. Sci. Res* **3** 1690-3

[13] Al-Hasan A Y and Ghoneim A A 2005 A new correlation between photovoltaic panel's efficiency and amount of sand dust accumulated on their surface *International Journal of Sustainable Energy* **24** 187-97

[14] Hussain A, Batra A and Pachauri R 2017 An experimental study on effect of dust on power loss in solar photovoltaic module *Renewables: Wind, Water, and Solar* **4** 9

[15] Kaldellis J and Kokala A 2010 Quantifying the decrease of the photovoltaic panels' energy yield due to phenomena of natural air pollution disposal *Energy* **35** 4862-9

[16] Appels R, Lefevre B, Herteleer B, Goverde H, Beerten A, Paesen R, De Medts K, Driesen J and Poortmans J 2013 Effect of soiling on photovoltaic modules *Solar energy* **96** 283-91

[17] Sayyah A, Horenstein M N and Mazumder M K 2014 Energy yield loss caused by dust deposition on photovoltaic panels *Solar Energy* **107** 576-604

[18] Elminir H K, Ghitas A E, Hamid R, El-Hussainy F, Beheary M and Abdel-Moneim K M 2006 Effect of dust on the transparent cover of solar collectors *Energy conversion and management* **47** 3192-203
[19] He G, Zhou C and Li Z 2011 Review of self-cleaning method for solar cell array *Procedia Engineering* **16** 640-5
[20] Gaier J R and Perez-Davis M E 1990 Aeolian removal of dust types from photovoltaic surfaces on Mars
[21] Gaier J R and Perez-Davis M E 1991 Effect of particle size of Martian dust on the degradation of photovoltaic cell performance
[22] Levy M and Levy E E 2010 Automated system for cleaning a plurality of solar panels. Google Patents)
[23] Jaradat M A, Tauseef M, Altaf Y, Saab R, Adel H, Yousuf N and Zurigat Y H 2015 A fully portable robot system for cleaning solar panels. In: *2015 10th International Symposium on Mechatronics and its Applications (ISMA)*: IEEE) pp 1-6
[24] Kawamoto H and Shibata T 2015 Electrostatic cleaning system for removal of sand from solar panels *Journal of Electrostatics* **73** 65-70
[25] Sayyah A, Horenstein M N, Mazumder M K and Ahmadi G 2016 Electrostatic force distribution on an electrodynamic screen *Journal of Electrostatics* **81** 24-36
[26] Zhang M, Feng S, Wang L and Zheng Y 2016 Lotus effect in wetting and self-cleaning *Biotribology* **5** 31-43
[27] Pan A, Lu H and Zhang L-Z 2019 Experimental Investigation of Dust Deposition Reduction on Solar Cell Covering Glass by Different Self-cleaning Coatings *Energy*
[28] Wang P, Xie J, Ni L, Wan L, Ou K, Zheng L and Sun K 2018 Reducing the effect of dust deposition on the generating efficiency of solar PV modules by super-hydrophobic films *Solar Energy* **169** 277-83
[29] Nguyen B Q H, Shanmugasundaram A, Hou T-F, Park J and Lee D-W 2019 Realizing the flexible and transparent highly-hydrophobic film through siloxane functionalized polyurethane-acrylate micro-pattern *Chemical Engineering Journal* **373** 68-77
[30] Li W, Tan X, Zhu J, Xiang P, Xiao T, Tian L, Yang A, Wang M and Chen X 2019 Broadband antireflective and superhydrophobic coatings for solar cells *Materials Today Energy* **12** 348-55
[31] Harun M H, Talib Z A, Ibrahim N A, Liew Ying Chyi J, Salleh N G N, Alias M S, Mohamed M and Othman N 2018 Characterization of Transparent Hydrophobic Coating with Silica and Graphene Oxide Fillers by Sol-Gel Method *International Journal of Nanoelectronics & Materials* **11**
[32] Park J, Park J and Lee D 2018 Optically transparent super-hydrophobic thin film fabricated by reusable polyurethane-acrylate (PUA) mold *Journal of Micromechanics and Microengineering* **28** 025004
[33] Saharudin K A, Sreekantan S, Basiron N, Chun L K, Kumaravel V, Abdullah T K and Ahmad Z A 2018 Improved super-hydrophobicity of eco-friendly coating from palm oil fuel ash (POFA) waste *Surface and Coatings Technology* **337** 126-35
[34] Beranek G D 2008 Solar Collector with Hydrophilic Photocatalytic Coated Protective Pane. Google Patents)
[35] Karimi A, Sohrabi B and Vaezi M 2018 Highly transparent, flexible and hydrophilic ZnS thin films prepared by a facile and environmentally friendly chemical bath deposition method *Thin Solid Films* **651** 97-110
[36] Mondal S, Mondal A K, Sharma A, Devalla V, Rana S, Kumar S and Pandey J K 2018 An overview of cleaning and prevention processes for enhancing efficiency of solar photovoltaic panels *CURRENT SCIENCE* **115** 1065
[37] Xue J, Wang Z, Hu W, Shen Q, Liu X and Jia H 2019 The surface wettability of TiO 2 nanotube arrays: which is more important—morphology or chemical composition? *Journal of Porous Materials* **26** 91-8
[38] Weihao L, Shengnan C, Jianxun W, Shaohui J, Chenlu S, Yong L and Gaorong H 2019 Study on structural, optical and hydrophilic properties of FTO/TiO2 tandem thin film prepared by aerosol-assisted chemical vapor deposition method *Surface and Coatings Technology* **358** 715-20
[39] Nayshevsky I, Xu Q and Lyons A M 2018 Hydrophobic–Hydrophilic Surfaces Exhibiting Dropwise Condensation for Anti-Soiling Applications *IEEE Journal of Photovoltaics* **9** 302-7
[40] Kivambe M M, Figgis B and Tabet N 2018 Performance of Antisoiling Coatings for Enhancement of PV Panel Performance in Doha. In: *Qatar Foundation Annual Research Conference Proceedings Volume 2018 Issue 1*: Hamad bin Khalifa University Press (HBKU Press)) p EEPD754

Cost-effective, Reliable, and Precise Surface Mount Device (SMD) on PCBs

B Iftikhar[1], M M Malik[1], S Hadi[1], O Wajid[1], M N Farooq[2], M M Rehman[1, 3] and A K Hassan[1]

[1]*Department of Electrical Engineering, Ghulam Ishaq Khan Institute, Swabi, Pakistan*
[2]*Cheif Executive Officer, SPUD Energy, Pakistan*
[3]*Author to whom any correspondence should be addressed.*

E-mail: muqeet@giki.edu.pk

Abstract. Automated Pick-and-Place (P&P) machines offer a new prospective for the assembly of Surface Mount Devices (SMDs) on Printed Circuit Boards (PCBs), owing to their advantages of time-effectiveness and precision. SMDs are widely used to improve production efficiency and reduce common defects. However, these machines are quite expensive and not easily available for the local market. This study is based on the design of a P&P machine model that incorporates principles of machine learning theory. The model has been developed in such a way that it can be introduced to the local market in a more cost-effective manner without compromising on its precision and accuracy.

1. Introduction

Assembly of components onto the Printed Circuit Board (PCB) uses mainly two techniques; the first, more prevalent method being the one that employs through-hole technology and the second, relatively advanced method employs Surface-Mount Technology (SMT). As no component leads must be inserted through holes in the PCB during assembly, Surface Mount Devices (SMDs) are well suited for high speed, reliable placement, miniaturization and low cost [1]. SMT facilitates large volume production by simplifying the integration of electronic circuits and simultaneously allows higher component density [2]. Due to the precision of the machine, small errors in component placement are corrected automatically as the surface tension of molten solder pulls components into alignment with solder pads.

The SMD assembly process can be classified in three categories [3]: Picking (the component is picked with a vacuum), Alignment (of the component relative to the board), and Placement. The accuracy with which components are to be placed on the PCB is a very crucial step in the assembly process [4]. Therefore, automatic machines, namely Pick and Place (P&P) machines, have been developed to precisely place various electronic components such as capacitors, resistors and Integrated Circuits (ICs) onto PCBs with high precision.

During the 1980s and 1990s two different P&P machines were placed in sequence for the assembly process. They were built around a single turret design where the they were used together for the placement of the components onto the board from a tape feeder. Owing to the huge cost of the two machines, their inflexibility and the speed limitations of the chip shooters, manufacturers shifted to an all-in-one modular, multi-headed and multi-gantry machine from the 20th century onwards. Nowadays, there are various kinds of SMD placement machines available, such as Sequential P&P, Rotary Disk Turret, Concurrent P&P etc. [5].

The P&P machines provide an easy and accurate method of assembling SMDs on PCBs but owing to their high cost, they are not affordable for industries. They must send their Gerber files for fabrication of PCBs to companies or markets that specifically employ these machines further increasing the cost and time required to get the desired product.

This study has been undertaken to fabricate a cost-effective P&P machine for the assembly of SMD without compromising precision and accuracy. The success of this project will have a huge impact on the growth of our local electronic industry.

2. Proposed Method

The initial steps of the design process included the study of automated commercial products of P&P machines already present in the market. Keeping in view the overall cost and time-effectiveness, it was concluded that the model design should be based on automatic P&P machines to cut the extra labour cost.

2.1. System Modelling

The system has been modelled as shown in Figure 1. The body of the machine comprises primarily of three axes, X, Y and Z, that are to be controlled using a MACH-3 controller connected to a computer. The X and Y axes have screw rods with timing belts that travel using Nema-17 stepper motors connected to wheels mounted on the timing belt. The Z axis is made of linear sliders mounted on the travelling plate of the X axis. Two Z axes are being used according to our design; one for paste dispensing and the other for the P&P action.

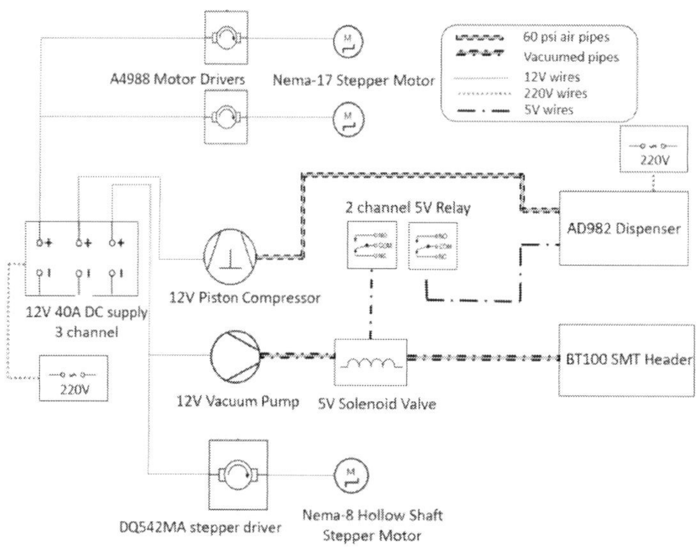

Figure 1. System Model of SMD assembly machine.

The P&P machine was programmed with the information about component positions, developed through Gerber File. The machine, after determining the position of a component through the program developed using computer vision, picks up electronic components, ICs, from the feeder tray (holds the components before they are placed onto the PCB). The ICs are picked through the vacuum nozzle connected to a hollow shaft, Nema-8 stepper motor through a header. Before the IC is placed onto its respective position on the PCB, a drop of high-quality solder paste is dispensed onto the corresponding location through a pressure control kit with its air outlet connected to a syringe.

The design for this model also includes two cameras mounted on the machine. One camera has been placed between the two Z axes to help with the precise placement of the components on the PCB. The other camera is faced upwards and has been installed on the platform. It is used to correct the

IOP Conf. Series: Materials Science and Engineering **899** (2020) 012007 doi:10.1088/1757-899X/899/1/012007

orientation of component in the case that it is not deposited in the correct position from the feeder tray. Both the cameras have been programmed using machine learning principles through OpenCV.

2.2. *Mathematical Modelling*

An important feature of the P&P machine, included in our model, is the paste dispensing module responsible for dispensing small drops of solder paste onto the PCB before an IC is placed onto it. It consists primarily the paste dispenser kit, AD982 and the 12V piston compressor (100 psi) with the final dispensing taking place through a syringe. High viscosity industrial solder paste has been used to better control the velocity and flow of the paste that exits the dispensing nozzle.

To increase accuracy of results, standard thermodynamic principles for the calculations of time have been made as in equation (1).

$$\dot{m} = \rho AV \; ; t = \frac{m}{\rho AV} \qquad (1)$$

where m is the mass of paste dispensed, ρ is the density of the solder paste, A is the area of the nozzle that dispenses paste, and V is the velocity of the that exits the nozzle.

2.3. *Algorithm Development*

The code for the microcontroller, object detection and scaling programs have been written and tested through OpenCV. Detection of the corner points and the centre of the IC was a crucial step for their precise placement on the PCB as seen in Figure 2. This was achieved using Algorithm 1:

Algorithm 1: Algorithm for IC Size Calculation

```
for c in cnts:
    if cv2.contourArea(c) < 100:
    then
    (tl, tr, br, bl) = box
    (tltrX, tltrY) = midpoint(tl, tr)
    (blbrX, blbrY) = midpoint(bl, br)
    (tlblX, tlblY) = midpoint(tl, bl)
    (trbrX, trbrY) = midpoint(tr, br)
    dA = dist.euclidean(tltrX, tltrY), (blbrX, blbrY)
    dB = dist.euclidean(tlblX, tlblY), (trbrX, trbrY)
    if pixelsPerMetric is None:
        pixelsPerMetric = dB / args[width]
    dimA = dA / pixelsPerMetric
    dimB = dB / pixelsPerMetric
end for
```

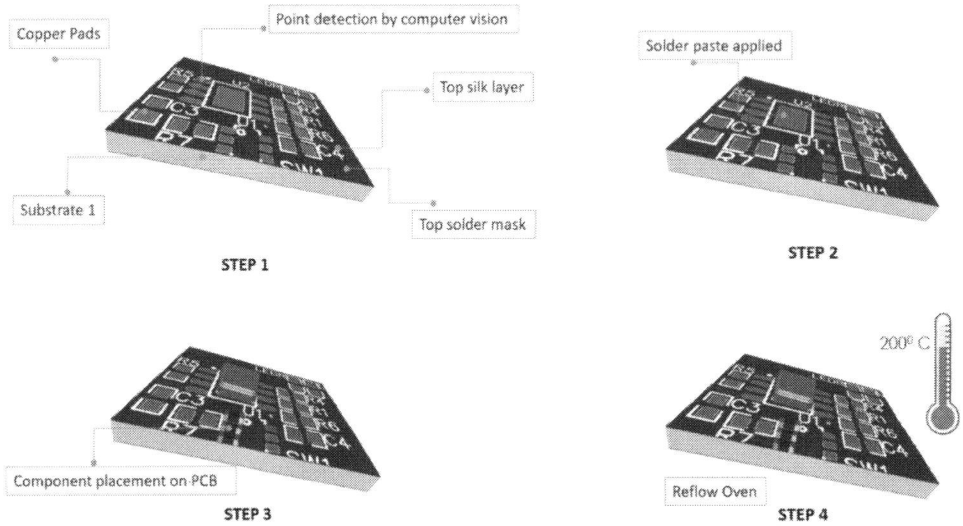

Figure 2. Schematic diagram for SMD assembly process.

3. Results and Analysis

We aim to bring a commercially available product to the local market, various design parameters were analysed and adjusted for the model of the P&P machine accordingly. The final design of the model has been curated after taking into consideration the challenges faced along the way.

The equipment required for the design includes a 12V, 40A and 3-channel DC supply to power the machine, Nema-17 stepper motors together with A4988 stepper drivers, a 12V piston compressor, an AD982 dispenser, a 12V vacuum pump to assist in the P&P action, a 5V solenoid valve, 2-channel 5V relay, a BT100 SMT header and Nema-8 hollow shaft stepper motor used together with a DQ542MA stepper driver to perform the picking action.

Initial testing of the paste dispensing module was conducted using a mechanical relay. However, after carrying out several tests, it was concluded that switching to a solid-state relay would make the process more efficient as it provided a faster switching speed (less than 400ms). The relay is controlled by the microcontroller with the help of an 8-inch jack that sends input signal to the dispenser via an aux cable.

The data used to plot the graphs in Figure 3 was obtained from a set of experiments carried out to establish relationships between parameters crucial to the design process. Fig. 3(a) represents a non-linear relation between the velocity of the paste dispensed and the switching time of the relay. As can be seen from the figure, change in switching time controls the velocity of the paste dispensed. Fig. 3(b) represents a linear relationship between the diameter of the dispensing nozzle and the volume of the paste dispensed. Choosing an effective nozzle group (or a nozzle) is important since a nozzle change operation is time consuming [6]. Different sets of nozzles were used during the experiment to determine which one provided the best control over the volume of the paste dispensed as this had to be very accurate. Greater or lesser volume dispensed could lead to inaccurate placement of ICs on the PCB. A decrease in nozzle diameter allows greater control over dispended volume.

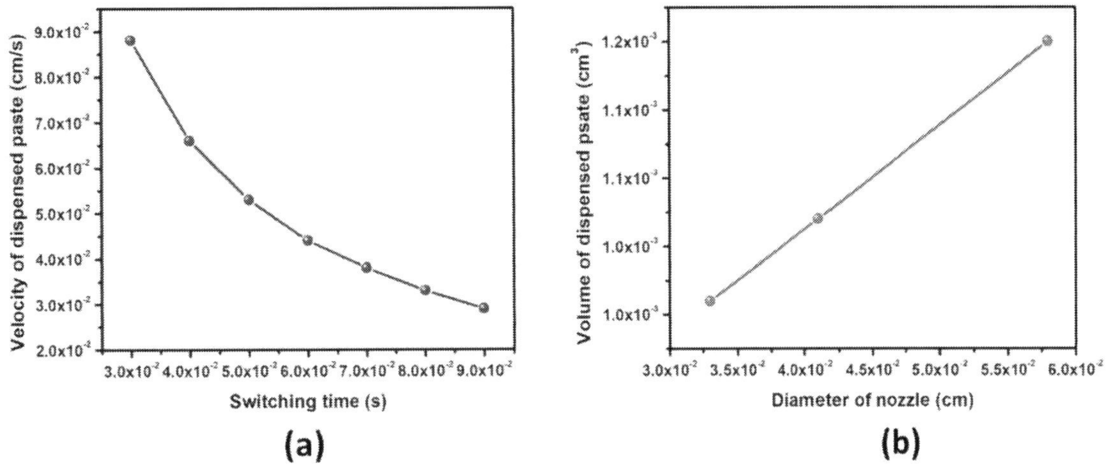

Figure 3. (a) depicts the relationship for the velocity of the paste dispensed and the time required; (b) depicts relationship of volume of paste dispensed and the diameter of nozzle.

4. Conclusion

With this study, we have proposed and implemented the design model for a P&P machine that operates on the principles of Machine Learning Theory. The model was completed within 18-20% of the cost of a commercially available P&P. With the study that was conducted, we have been able to reduce the cost factor and provide an alternate solution that can be introduced to the local market without compromising on the accuracy and precision required to complete the assembly process.

Acknowledgement

The group members would like to thank our sponsor SPUD Energy for their generous funding.

References

[1] G. Burel, F. Bernard and W.J. Venema, "Vision Feedback for SMD placement using neural networks," Proceeding IEEE, pp.1491, 1995.
[2] Lifei Bai, Xianqiang Yang and Huijun Gao, "Corner point-based coarse-fine method for surface mount component positioning," IEEE Transactions on Industrial Informatics, vol.14, issue 3, March 2018.
[3] Ivo Van RIJSSEL, "User requirements of the Galatea vision work package dedicated to SMD placement", Galatea Report (European Esprit project no 5293), April 15th, 1991.
[4] Huanca, J. and Spence, R." A new statistical algorithm for fault location in tolerance analogue circuits," Proceedings IEE, vol. 30, part G, pp. 1148-1151, 1983.
[5] Gastel, 2002; Grotzinger, 1992; Khoo and Loh, 2000.
[6] M.Ayob and G. Kendall, "A nozzle selection heuristic to optimize the hybrid pick and place machine," IEEE Conference on Cybernetics and Intelligent Systems, pp.1260, 05 July 2005.

Solar Thermal Opportunities and Challenges in Pakistan

Adnan Ayaz[1], Faraz Ahmad[2*], Saim Saher[3], Muhammad Noman[4]

[1, 4]U.S.-Pakistan Center for Advanced Studies in Energy, University of Engineering and Technology, Peshawar, KPK, Pakistan

[2]Department of Mechanical and Aerospace Engineering, Air University Islamabad, E-9, Islamabad Capital Territory, 44000, Pakistan

[3]CECOS University of IT and Emerging Sciences, Peshawar, KPK, Pakistan

[*]**Corresponding author:** faraz.ahmad0460@gmail.com

Abstract. In June 2018, the power deficit in Pakistan hit a record high of 9000 MW. Although alarming, it is not unprecedented due to past trends of exponential population growth, inadequacy of conventional energy sources to meet demand and lack of technological advancements. This represents an urgent and unavoidable need to turn to alternate renewable energy sources such as solar, wind and hydroelectric power. This paper aims to explore the possibilities presented by solar energy to solve the power crisis. This technology can be used to generate electricity which can then be used in different thermal applications. This paper presents a detailed discussion regarding these technologies, the institutions working in this field and scheme of solar thermal plants as per energy needs of Pakistan.

Index Terms: *Solar energy, conventional energy, solar collector, solar energy potential.*

1. Introduction

In recent times, frequent increase of population has been observed worldwide, particularly in developing countries with weak economies. This increase in population is inextricably linked with an increasing energy deficit. Since the economy and energy demand of a country are closely linked, as the economy of these developing countries grow, so does their energy demand. In the span of 2006 to 2030, marketed energy consumption of the whole world has been projected to be increased by 44% which is quite a brisk increase in the demand of energy globally [1]. This increased demand for energy can be catered only by introducing renewable resources of energy. These renewable sources of energy are cost effective and sustainable solution to energy crisis faced by developing nations.

The sun is considered to be one of the best alternative sources of energy. Due to its high surface temperature, it emits rays which are received by earth as form of solar energy. Being a natural emitter, the sun provides continuous yet varying amounts of energy to the earth, subject to the orientation of each area towards sun. Surfaces which directly face the sun receive more solar radiation and vice versa. As solar energy is renewable, it does not produce harmful effects to the environment.

Past studies show that fossil fuels provide 88% of the total energy which is consumed for all purposes across the globe. Fossil fuels are harmful for the nature and affect environment in several ways [2]. Greenhouse gases emitted by burning fossil fuels significantly contribute to global warming. 1.6 billion tons of carbon was emitted in 1950 alone due to the burning of fossil fuels across the globe [3]. Burning of fossil fuels negatively impacts the environment as well as human lives. Every year, almost 160,000 people expire due to adverse effects of climatic changes which may increase twofold by 2020 [4]. Incorporation of solar energy for the production of electricity limits the use of fossil fuels, thereby contributing to control global warming. Solar energy is also renewable and sustainable, unlike fossil fuels which are depleting rapidly. This characteristic makes solar energy more appealing as a solution to combat the energy crisis.

Sun provides 100,000 TW energy to the surface of earth, which when extracted for an hour, is considered to be sufficient to fulfill energy needs of all humans for a year [5]. Proper harnessing of such a phenomenal amount of energy can be the answer to the current, and oncoming energy deficit.

Pakistan is an energy deficit country with a shortfall of several thousand megawatts. In 2008, the shortfall of the country stood at 4500 MW [6]. There are many remote areas throughout the country which are still not connected to national electricity grid. This is mainly due to the exponentially increasing population of the country, the unstable economy and government as well as poor infrastructure. Another key factor is the constantly fluctuating oil prices which make oil import for electricity generation more expensive. However, Pakistan has got abundant solar energy potential which can be harnessed in multiple ways to generate electricity for the country.

Solar energy can be harnessed in many ways. One way of extracting solar energy is to capture it through solar collectors. Solar collectors have the capacity to capture maximum possible solar energy and transfer it to a working fluid. This heated working fluid can be used for domestic heating purposes directly or stored for future use [7]. Solar collectors are also used to produce steam by concentrating sun rays on a particular structure containing water. Once this water is heated, it is converted to steam which is used to drive the rotor of a steam turbine. This is practiced worldwide to get extract maximum energy. These systems are especially useful in regions where there is a large number of sunshine hours.

Considering the entire situation, solar thermal generation seems to be a suitable and feasible mechanism to cope with the increasing energy demand in Pakistan. This clean technology ensures electricity generation with almost no adverse environmental effects. It is currently one of the best choices to mitigate the power crisis in the country. This paper aims to highlight the various advantages, hurdles and possibilities regarding solar thermal generation in Pakistan.

2. Solar Energy Potential of Pakistan

The latitude of Pakistan is between 620 and 750 East while its longitude falls under the range of 240 and 370 North [8]. Being in the sunny belt, it has been found to have tremendous solar energy potential standing at 2.9 TW [9].

The geography of the country is such that sunlight is evenly and plentifully distributed throughout the country. Average irradiation received by a flat surface in Pakistan has been estimated to be 200 to 250 watts per m^2, which is sufficient to make the country self-dependent in terms of power demand [10]. Sun shines in the country for almost 300 days annually. The topography of the country allows it to harness 1.9 MJ to 2.2 MJ solar energy. Many regions across the country have got enormous solar energy potential, Baluchistan being at the top. Baluchistan has such great potential that it is counted as one of the richest regions regarding solar energy potential. This claim is supported by the fact that on a particular sunny day, which comprises of 8 to 8.5 hours on average, Baluchistan receives irradiation of 19 to 20 MJ/m^2 [2].

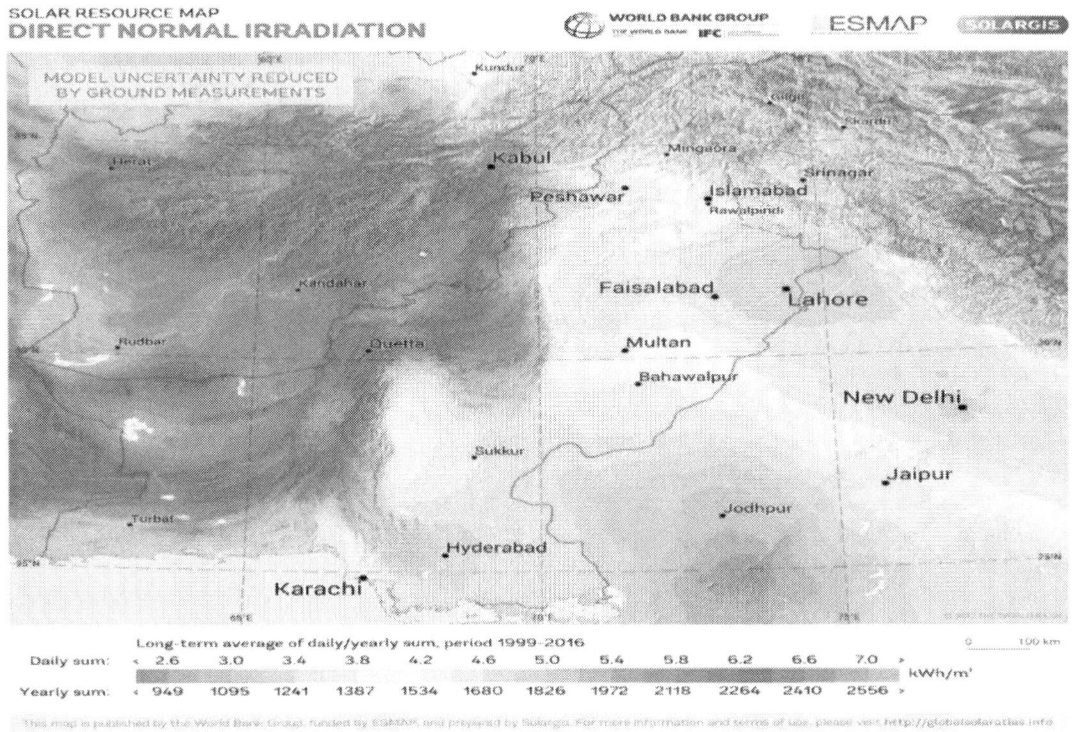

Figure 1. Solar insolation map of Pakistan

3. Solar Thermal Technologies

The working principle of this technology it to harness solar energy to heat a working fluid, which is then utilized for various purposes. Normally two types of thermal power systems exist: Non-Concentrating Solar Systems and Concentrating Solar Power Systems (CSPs).

3.1 Non-Concentrating Solar Systems

These differ mostly because non-concentrating solar systems produce less temperature since solar energy is not focused on a particular point. These systems have higher efficiencies, moderate costs and are normally suitable for intense climates. These properties make them suitable for domestic purposes. As of 2010, this technology has benefited above 70 million residential areas globally [1]. Two forms of this technology are widely used these days: Flat Plate Solar Collectors and Vacuum Tube Solar Collectors.

3.1.1 Flat Plat Solar Collectors

Flat plate collectors are made up of multiple parts which are described as follows.

3.1.1.1 Absorber Plate

Absorber plates are made up of such materials which enable them to capture maximum amount of solar radiation. Sometimes, selective coating is also applied over the surface of the absorber plate to enhance its absorbing capacity.

IOP Conf. Series: Materials Science and Engineering **899** (2020) 012008 doi:10.1088/1757-899X/899/1/012008

3.1.1.2 Transparent Glass Sheet

A Transparent Glass Sheet is provided on top of the absorber plate to reduce heat loss from absorber surface.

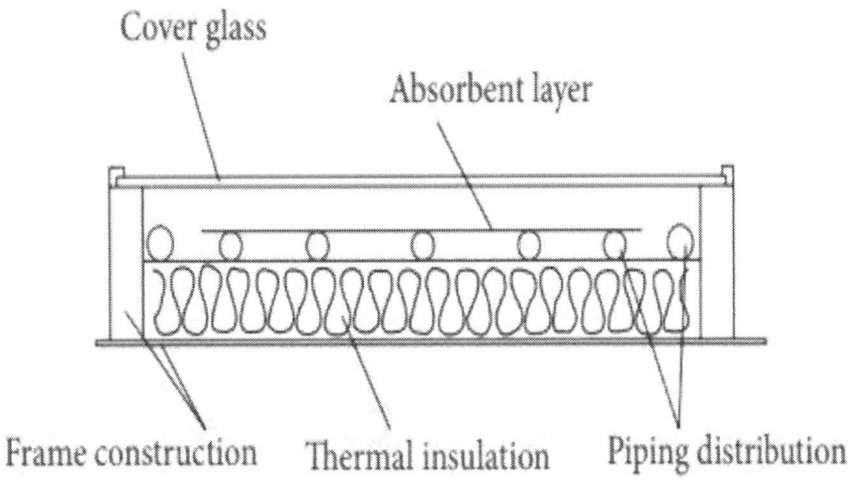

Figure 2. Flat plate solar collector [10]

3.1.1.3 Metal Pipes

Metal pipes are directly attached to the absorber plate to enable maximum heat exchange with the working fluid present within the metal tubes.

Further, an insulated support is installed to minimize heat loss from the surface of the absorber plate. All parts are then protected by an external shell [11]. Since Flat Plate Collectors absorb solar radiation from any direction, the need for a complicated tracking system (required to appropriately position the collector) is eliminated.

3.1.2 Evacuated Tube Solar Collectors

Evacuated tube solar collectors are generally better in performance because they produce higher temperatures than their flat plate counterparts. It consists of following parts.

3.1.2.1 Vacuum Glass Tubes

Vacuum filled glass tubes reduce convective and radiation heat loss.

3.1.2.2 Copper Tubes

These are used to transfer solar energy to the working fluid in the form of heat.

An outer casing made of aluminum protects the collector from the influence of the external environment, thus enabling it to be used in harsh environments. However, this type of collector requires a high capital cost [11].

IOP Conf. Series: Materials Science and Engineering **899** (2020) 012008 doi:10.1088/1757-899X/899/1/012008

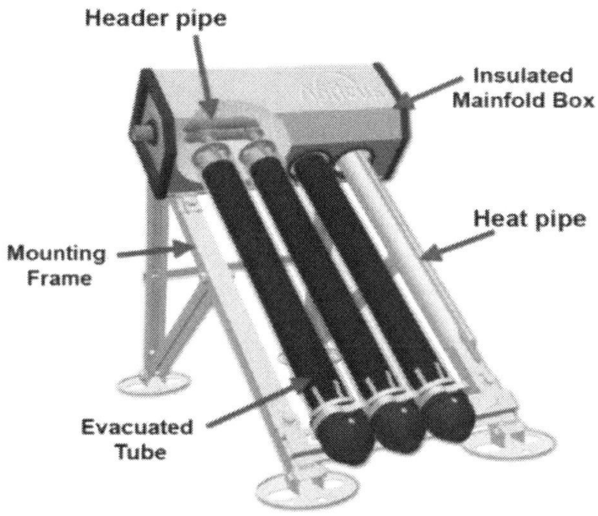

Figure 3. Evacuated tube solar collector [12]

3.2 Concentrating Solar Power Systems (CSPs)

CSPs are more common and used globally for power generation. Its working principle is similar to that of steam turbines. Solar radiation is directed to solar collectors or mirrors to concentrate it on a particular structure. This structure contains a working fluid which is heated to produce steam which is used to run the rotor of a steam turbine to generate power [13]. There are two types of CSPs [14]; Line focusing CSPs and Point focusing CSPs.

3.2.1 Line focusing CSPs

Line focusing solar thermal collectors and power systems focus sun rays on a line where working fluid is heated. In these power systems, sun rays are reflected from a surface in such a manner that they are focused in the shape of a line on a structure where working fluid gets heated to carry out the desired function. Line focusing solar collectors can achieve temperatures in the range of 350 °C to 550 °C [15].

3.2.2 Point focusing CSPs

Points focusing solar collectors are those which, when exposed to direct solar radiations, focus all radiations on a single point to heat a working fluid. This heated fluid then performs a specific function, usually power generation.

There are different forms of CSPs through which higher temperature of particular fluids can be achieved for various purposes. Past studies show that multiple CSPs have been made to achieve the best possible conversion of solar energy to heat energy. Today, many choices are available to collect solar energy and convert it into heat energy with best possible efficiency. As technology advances, many changes are observed on a regular basis to modify the structure of different collectors to absorb maximum solar energy.

Pakistan has not yet moved to produce power using CSPs [16]. The following four collectors [15] are very important and can produce sufficient power for Pakistan to mitigate its blackouts: Parabolic Troughs, Parabolic Dishes, Solar Tower and Linear Fresnel Reflectors.

3.3 Parabolic Troughs

Parabolic troughs are widely used to generate power in energy deficient areas. They are a type of line focusing CSPs. This technology uses a parabolic trough which has the distinctive property of concentrating sun rays over a receiver containing the working fluid. The receiver is a tube, coated with a specific material, receives concentrated solar rays reflecting from parabolic trough. Sometimes, the receiver tube is fitted in evacuated glass to minimize heat loss. It is strategically placed in the optical focal line of the parabolic trough for maximum efficiency. To further improve efficiency, one dimensional tracking systems are also installed to enable them to capture maximum direct radiations. Without tracking, solar rays may not always be directed perpendicular to the surface of the trough, which reduces the intensity and the temperature in the receiver.

In parabolic troughs, the fluid inside receiver can be heated up to 400 °C, which is then used to produce steam. This steam then runs the rotor of turbine to produce the required power [17].

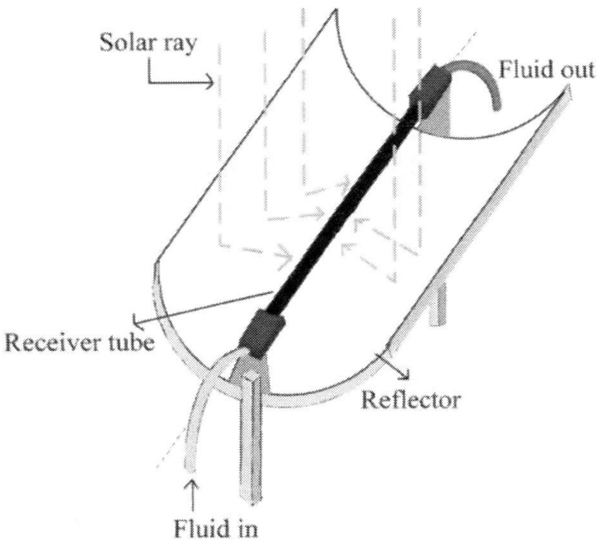

Figure 4. Parabolic trough collector [18]

3.4 Parabolic Dishes

Parabolic dishes are point focusing solar collectors which capture solar rays and direct them to a point to convert this solar energy to heat energy. These solar collectors can achieve a very high temperature of 1500 °C [19].

Parabolic dishes are in the shape of a paraboloid, which is made by revolving a parabola about its own axis. The surface of the dish, which directly faces sun, is made up of highly reflective material to reflect and concentrate maximum solar radiation. The receiver, which again contains the working fluid, is installed at the focal point of the parabolic dish. It is generally made up of a metal for maximum heat transfer to the working fluid. It is tightly attached to a support which mounts it in a fixed position. The working fluid is heated up to the desired temperature and then utilized either in Stirling engines or steam power plants to produce power.

Parabolic dishes also need proper two-dimensional tracking systems to face the sun directly to receive maximum solar radiation. Since the parabolic dish is rotated in two dimensions, it requires extra strength. Hence, a foundation along with supporting frames are also installed in the system [20].

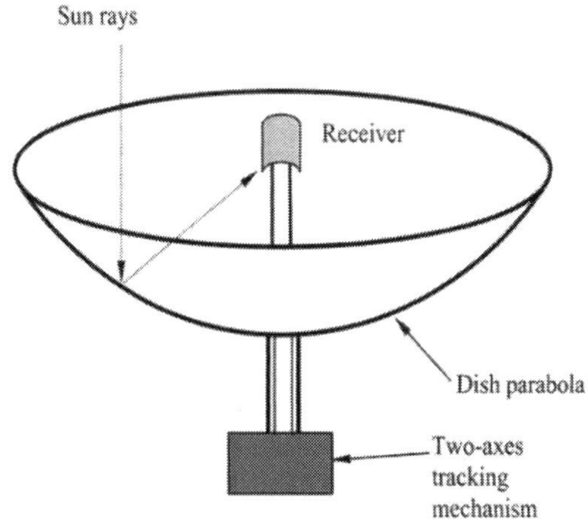

Figure 5. Parabolic dish collector [21]

3.5 Solar Tower

Solar tower is one of the best options to harness solar energy for power generation. This technology consists of a central receiver system (CRS) and widely distributed mirrors. These mirrors are called heliostats. Each heliostat tracks sunlight to focus it on a receiver mounted at the top of the tower. Working fluid is present inside a heat exchanger which is heated up to a certain temperature to serve a desired function [15]. The working fluid inside the receiver may be water, air or molten salt [20]. Generally, a cylindrical receiver is mounted to receive maximum reflected rays from the heliostats. When solar rays are concentrated on a particular CRS, the working fluid is heated, which in turn, is stored in a thermal energy storage tank or used directly in power generation systems. Thermal energy stored in a tank may be used to generate power during the night or on cloudy days. Solar towers can achieve extremely high temperature up to 1500 °C [22].

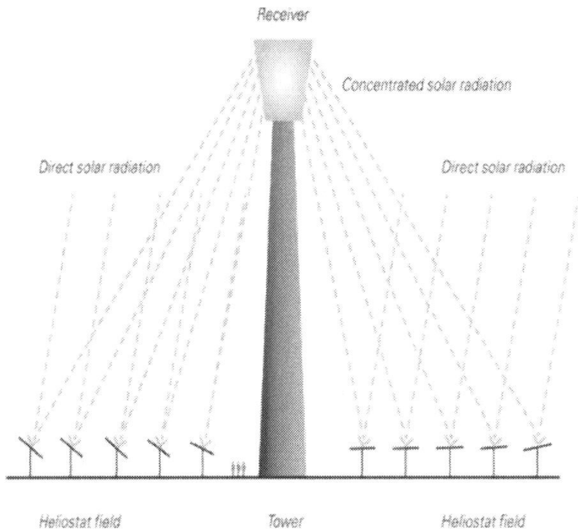

Figure 6. Solar tower [23]

3.6 Linear Fresnel Reflector

Linear Fresnel reflectors also contain mirrors which are used to focus solar rays on the receiver. In 1993, the first Fresnel Reflector was developed in the University of Sydney, and later patented in 1995. When sun rays fall on the surface of the mirrors, they are focused on a linear metal receiver. Since this utilizes the line focusing technique, a single axis tracker is required to directly face the sun. The receiver tube contains the working fluid, which is heated by concentrated solar radiations. This highly heated working fluid can be used either directly used for power generation or passed through heat exchangers for steam production, which is then used to operate steam power plant [22].

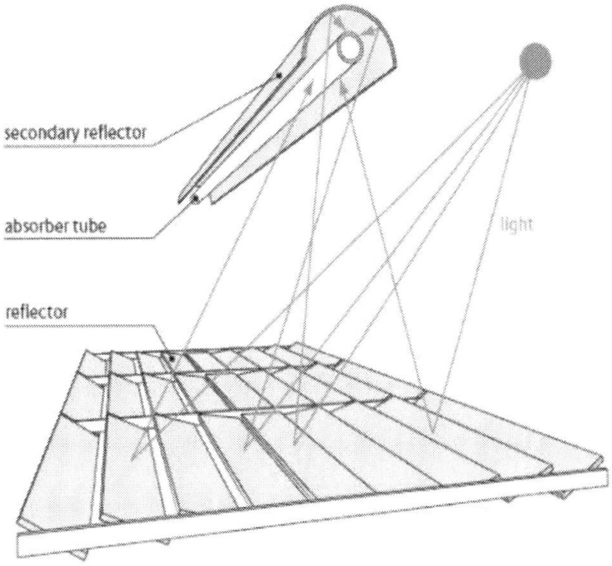

Figure 7. Linear Fresnel Reflector [24]

4. Comparison of Different Concentrating Technologies

There are various concentrating technologies available worldwide for power production. Some are point focusing while others are line focusing. Each technology produces different temperature ranges depending upon the collector's surface. Those surfaces which concentrate more light rays, produce greater temperature and vice versa. Table 1 compares various features of these concentrating technologies.

Table 1. Comparison of Different Concentrating Technologies [23]

Technology	Parabolic Troughs	Parabolic Dishes	Solar Tower	Linear Fresnel Reflector
Temperature	400 °C	750 °C	1000 °C	270 °C
Efficiency	10-15%	18-25%	14-17%	9-15%
Concentration	Line	Point	Point	Line
Cost ($/KW)	4200	6000	4500	2200

5. Working of a Typical Solar Power Plant

Solar thermal power plants operate on the basis of the Rankine cycle. The Rankine cycle incorporates various components of the power plant. A solar thermal collector functions as an evaporator where water is heated to produce steam. This high pressure steam is used to turn the rotor of the turbine. A condenser then condenses water under low pressure. Water is then pumped to the evaporator and the cycle is repeated[25].

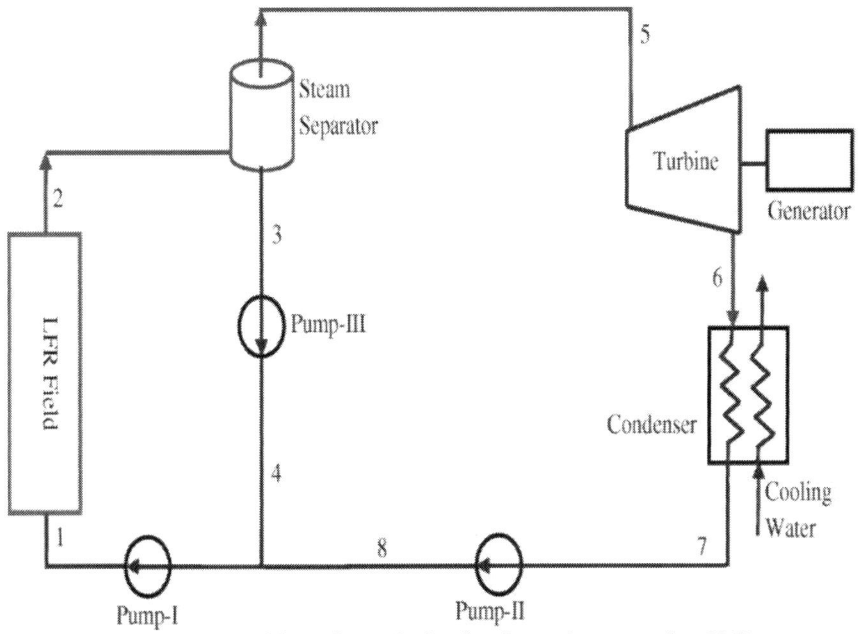

Figure 8. Working of a typical solar thermal power plant[26]

6. Solar Thermal Applications is Pakistan

As established before, Pakistan has high solar energy potential due to its advantageous geographical position. While this energy can be utilized to generate power, it can also be used for other purposes. Some of these alternative applications are discussed below.

6.1 Solar Water Heating

Solar water heating (SWH) is an efficient, globally used technique to heat water with the help of solar energy. Solar energy, collected by a solar collector, is converted to heat energy and transferred to water thereby increasing its temperature. This is a rapidly emerging and growing technology, currently implemented in multiple countries, due to its efficient results. Solar water heating can be used for several domestic purposes, commercial activities and industrial processes. They can also be used for air conditioning through dedicated absorption chillers [27]

This technology is especially relevant because water heating consumes a great amount of energy globally. It has been found that 15 to 20% of the total domestic energy consumption is used to heat water. The United Kingdom and the United States of America have been found to be using 23% and 18% of their total domestic energy to heat water respectively. As in domestic applications, water heating is also of prime importance in the industrial sector. Heated water is used for various processes carried out in different industries. The textile industry consumes 65% of its energy to heat water for different purposes. Past studies show that Pakistan uses 10% of its domestic energy to heat water [6]. SWHs are also preferable because they do not emit carbon unlike fossil fuels.

Although, these systems are costly [27], their payback period seldom exceeds three years [6]. Solar water heating technology is gaining popularity in northern part of the country because of the cold environment and lack of availability of natural gas [28].

As the population of Pakistan increases, the energy demand for domestic and industrial purposes is increasing as well. A large portion of this energy is used to heat water. The textile industry of Pakistan provides 60% of the total income earned due to export. At the same time, it needs water heated up to 80 °C for various purposes [28]. This makes solar water heating a viable option for domestic and industrial needs in Pakistan.

6.2 Solar Cookers

Cooking is one of the most vital phenomena occurring worldwide. Often, it consumes a lot of heat energy since the food being cooked contains a lot of water. South Asian countries are considered to be among prime users when it comes to energy consumption for cooking purposes. They consume a great portion of primary energy for cooking which amounts to one third of the consumption of primary energy [16]. Water has a boiling point of 100 0C. In order to completely remove water from the food, temperatures between 90 and 100 0C must be supplied. When food is cooked, more energy is required for sensible heating. Thus, while cooking food, 20% of the heat is used to boil it, 35% is used to remove water and the remaining 45% is lost to the external environment in the form of heat losses [29].

Solar cookers are one of the feasible options for cooking in the countries like Pakistan. Solar cookers are relatively cheaper than wood or other fossil fuels which makes them an attractive option for people with limited financial resources. These cookers also reduce the wood utilized for cooking, thus decreasing deforestation.

There are two types of solar cookers available today: Box type solar cookers and Concentrator type solar cookers. Box type solar cookers use direct solar radiations for cooking purposes whereas concentrator type solar cookers have a solar collector, which concentrate solar radiations to heat a working fluid, which is then used for cooking.

Pakistan is quickly adopting solar cooker technology. Many public, private and non-governmental organizations are engaged to develop this efficient and low cost technology. South Asian countries like China and India have already adopted this technology and are getting great advantage through it. China is quite familiar with this technology. 60,000 and 14,500 solar cookers are used across China and India respectively. Pakistan, when compared to these countries, is quite far behind adopting this technology because only 5000 of these cookers are present in the country. Food cooked in solar cookers is healthy because solar energy does not impact the amount of vitamins in the food. Moreover, solar energy does not alter taste and flavour of the food [28].

There are certain barriers which prevent the widespread use of this technology throughout the country. One of the main causes is slow cooking. Food cooked in solar cookers takes more time compared to conventional cookers. Solar radiations are not always directed perpendicularly, thus solar tracking is required to achieve greater efficiency. Another issue faced by this technology is that it produces lower temperatures compared to conventional cooking temperatures. Sunshine hours also play a critical role. Solar radiations are unavailable for more than 60% of the day which makes solar cooking impossible for a greater part of the day [16]. Furthermore, people are still not mentally prepared to accept this technology as an alternative to conventional cooking. Public awareness through various campaigns and promotional ventures are required to educate people about this technology. Since the population in cities has increased exponentially, there is limited space available in urban areas which further poses a hindrance to the implementation of this technology. Initial cost of solar cookers is also quite high which does not allow poor people to take advantage of this technology [30].

6.3 Solar Dryers

Pakistan is an agricultural country which produces a lot of products including vegetables and fruits. Fruits are a vital part of the country's economy. Fruits are either consumed while fresh or dried for future use in off seasons. Solar energy is crucial in this aspect because it is considered to be the best choice for drying these fruits. Two methods are commonly used to dry these fruits in Pakistan: direct solar drying and convective solar drying. Direct solar drying is not a preferable technique because fruits are directly exposed to solar radiations in open atmosphere which makes them vulnerable to dust, grimes, insects, birds, animals and harsh environmental conditions. Moreover, this type of drying leads to degradation of the fruit [28].

Crops and fruits demand intensive care regarding drying. They need to be properly handled during this period to retain their natural taste and flavour. Solar drying, unlike conventional drying, has proved to be a better choice for drying agricultural products, like fruits, because it does not deteriorate their quality and retains their natural composition. Food products being dried are highly sensitive to high temperatures. They need to be provided with suitable temperature ranges. While open atmosphere solar drying is highly recommended for fruits, it is less preferable for vegetables, which usually get rotten due to high temperature of solar radiations [31].

Convective solar drying has proved to be better choice compared to open atmosphere drying because it is smoother, faster and more efficient. The low manufacturing and operating costs of solar dryers also make them a feasible option for drying different products [1].

Solar dryers are gaining popularity in the mountainous regions of Pakistan e.g. Gilgit Baltistan, which produces a vast variety of fruits. Apricots are abundantly found in this region. However, this region lacks basic infrastructure due to which a huge amount of fruit is spoiled and wasted every season, leading to a major economic loss. Thus, solar dryers are a suitable option for drying and preserving apricots. Although people are familiar with solar drying technology, many non-governmental organizations are working hard to create public awareness [32]. Punjab and Sindh can also utilize their solar energy potential for solar drying thereby creating opportunities for local residents [33].

The energy required for solar drying of a product depends on the moisture content of the product before and after drying. Different products require different temperatures for drying. Maximum energy consumption occurs when water evaporates since this process requires a large amount of energy. Relative humidity and air velocity also play very important role in solar drying [34].

Solar dryers are being used across the world to preserve different sort of fruits, vegetables and other agricultural products. These dryers have produced good results and are economically viable. They are also attractive since their functioning does not harm environment. Drying fruits and crops through this technology improves the standard of the product and increases their shelf life. It also reduces the consumption of fossil fuels.

Currently, different sorts of solar dryers are available in market depending on the shape, size and design. The needs of a customer determine the appropriate solar dryer required. This technology can also act as a viable option for food industries to preserve high quality food. By using this technology, solar energy and hence renewable energy will be familiarized, leading to reduction in the usage of conventional fuels and thereby decreasing environmental pollution [33].

Pakistan is a country which produces large amount of fruits, crops, vegetables and other agricultural products. A large amount of these products is wasted every year due to low demand and lack of storage capacity. People who grow these products cannot get maximum benefit due to wastage and hence receive less return on investment. In Pakistan, open air drying is more common compared to solar convectional drying which results in lower quality product and larger wastage. Open air drying is also less favourable because products are open to dust, insects, wind and rain. When dust gets stuck to the products, it can cause diseases in consumers. Thus, solar drying is preferable since it is reliable, does not depend on conventional energy resources, reduces electricity bills and is environment friendly.

6.4 Solar Desalination

Desalination is the process of removal of salt from brackish or seawater to make it fit for different purposes. Solar desalination uses solar energy for desalination of brackish water [28]. Water beneath the earth's surface contains dissolved salt which makes it unfit for drinking and laundry purposes. Usage of this water is also dangerous for health [32]. Many areas of Baluchistan, Sindh and southern Punjab suffer from lack of availability of sweet water. Solar desalination technology can be an option for these far flung areas because it is operationally simple and economical. It also reduces the number of germs present in water. So far two plants have been brought into operation. These purify 6000 gallons of water each day in Gwadar, Baluchistan. This project has facilitated many people in the area who are now taking advantage to purify their drinking water through this project [8].

Non-availability of sweet water is one of the main challenges in Cholistan and Thar deserts. Inhabitants of these deserts do not have access to fresh water due to its scarcity which compels them to use brackish water. Very few areas in these deserts contain suitable drinking water whereas the rest contain brackish water [35]. Salinity is also one of the major problems across the coastline of Baluchistan where the presence of dissolved salts renders the water used for multiple domestic purposes [36].

As discussed above, solar desalination uses solar energy to remove salts from the water. Two methods can be used to desalinate water with the help of solar energy. First method of solar desalination is called solar distillation. In solar distillation, water is heated with the help of solar energy and converted into vapours. These vapours, which are free from salts, are collected and condensed. Another method of solar desalination converts solar energy to thermal or electrical energy, which in turn is used to power conventional desalination methods. This is termed solar driven desalination. Research studies carried out at Awania village of India show that solar distillation is a more viable option for off-grid remote areas with fewer communities [37].

An easy method used for desalination of water uses the water still. The water still is comprised of a tray and a glass cover. Saline water is poured in a shallow tray. This tray is kept inside a glass cover with an inclined surface. When solar radiations fall on saline water, it is heated. Due to this heat, water evaporates. These vapours condense when they touch the relatively cooler glass surface. This fresh liquid water is collected at the bottom and can be used for various purposes. Through this mechanism, that 4 kg/m^2 of desalinized water can be produced daily, given that there are enough solar radiations [9].

Water stills are used worldwide to desalinate water. Pakistan is also a country which is quickly adopting this technology with different organizations working on making efficient solar stills. The Pakistan Institute of Engineering and Applied Sciences (PIEAS) is one such organization which has worked on the fabrication of solar stills. It has fabricated an optimized glass cover single basin solar still which showed an efficiency of 30.56%, which can compete with the efficiency of solar stills produced globally. This solar still is also economically competitive and can benefit poor people who live in remote areas [9].

6.5 Solar Water Pump

Pakistan has many areas where water is scarce. Water scarcity occurs due scarce and unpredictable rainfalls throughout the country. The Cholistan desert is one such area where water is scarce, relative humidity is low and evaporation and wind velocities are high. However, the solar intensity is also high in this particular desert. Other regions of the country, mainly many parts of Punjab and some part of Sindh also suffer from a similar problem. People store rain water and use it for different purposes. On hot summer days, when there is no rainfall, existing water evaporates leaving dry areas. When there is a severe water crisis, people migrate to other regions of the country, along with their domestic. These people live a very low standard of life due to lack of facilities. This is a serious injustice, which prevents them from taking part in activities which help the country grow. This makes it necessary to provide these people with opportunities to make their lives easier [9].

Solar water pumps can be used to pump water in these areas. Water can be pumped from beneath the surface of earth towards its surface by using a 1 or 2 hp engine, which is quite effective. Many organizations, like Pakistan Atomic Energy Commission (PAEC), are working to develop solar pump technology in the country. PAEC has developed a solar water pump which runs through a 2 hp steam engine. Steam is produced through a boiler with the help of 21 feet diameter paraboloid which contains flat mirrors [35]. The steam produced operates a steam engine, which in turn is used to run a solar water pump. Solar pumps can also be operated through PV modules. People related to farming have used these PV pumps in different regions of the country. The Agricultural Development Bank of Pakistan (ADBP) distributed these pumps among users [9].

Solar pumps can be a revolutionary technology, but at the moment, its initial capital cost prevents it from being implemented throughout the country. China, a technologically advanced country, required 261 MW in 2010 for solar water pumping [38]. Baluchistan can prove to be a huge beneficiary of the solar pump technology, if Pakistan learns from China.

6.6 Solar Ponds
Solar ponds are specifically built for collecting and storing solar energy. These ponds work on the principle of not allowing convective heat transfer by maintaining high temperatures at the bottom of the pond. To prevent convection, salt also is mixed in water. Such ponds are called salt gradient solar ponds. This technology is gaining fame and multiple solar ponds have been constructed across different countries over the last few years. The hot water can be used for different purposes e.g. generating power and desalination. Moreover, it can also be used for agricultural purposes and in dairy farms [39]. Like other countries, solar ponds technology can be a useful option for different applications in Pakistan.

7. Possibility of installation of Solar Power Plants in Pakistan
In Pakistan, cheap and sustainable sources of power generation are the need of the hour. Federal and provincial governments agree that new power plants are required to cope with the increasing energy demand. Power plants, particularly solar power plants, can fulfil this need on a national level. The Sindh government has signed a Memorandum of Understanding (MoU) with one of the leading companies in Germany, Azur solar. According to this MoU, a 50 MW solar power plant will be constructed in a village called Dhabeji in Thatta. The company will start with building a 60 kW power station. This power station will benefit the residents of the undeveloped villages, educational institutions like schools and other fundamental health care centres in the area of Badin [40].

A 50 MW solar thermal power plant in Jacobabad can be used to assess the viability of solar thermal power plants in Pakistan. This power plant uses the Stirling dish technology to produce electricity. Prior to constructing a solar thermal power plant, reliable solar insolation data is required. In case of this power plant, data was provided by a well-established organization of the country, the Bureau of Meteorology (BoM). Solar insolation data indicates that Jacobabad is among those regions which show higher solar energy potential. Solar energy potential of this region is sufficient, with the minimum solar energy potential being 4.45 kwh/m^2/day. This is quite high compared to the average solar energy potential of the world, which is 3.61 kwh/m^2/day. This value of solar energy potential suggests that it can be harnessed even when there is diffused irradiation and low sun intensity [40].

Once solar data was available, the economic and technical feasibility of this project was checked. A 25 kW Stirling dish system was used for this analysis. Dish technology is considered a popular option because it can harness solar potentials as less as 2 kWh/m^2 and can be used on cloudy days as well. Jacobabad, situated in Sindh, experiences extreme temperatures and low rainfall. The temperature in this region ranges from 45 to 50 ^{o}C in summer and 10 to 25 ^{o}C in winter. The low temperature average is 19.95 ^{o}C and high temperature average is 34.15 ^{o}C with overall average temperature being 27.05 ^{o}C [9]. This makes it quite similar to a desert, where global irradiation is as high as 1735 kWh/m^2/year [41]. The land required for installation of solar power plant was available at a cost much lower than the cost required for acquisition of land in United States of America and Europe [42].

Solar thermal power plants can be a good move for the mitigation of the energy crises in Pakistan. The 25 kW dish technology can act as a suitable source of harnessing solar energy for electricity production. This technology can be appropriate for electricity production in multiple high solar potential regions of the country like Federally Administered Tribal Areas (FATA), Sindh and Baluchistan. In Sindh, Jacobabad is ideal for the installation of solar thermal power plants operated via dish technology because its solar energy potential is more than 1700 kWh/m^2/year [40].

When it comes to economic considerations, the maintenance and operation costs of solar power plants ($0.001/kWh) are higher compared to conventional power plants. However, these costs are acceptable if the government is keen to invest in renewable energy. If subsidies given to conventional power plants are given to solar power plants, they can be quite beneficial for the future of the country. Other calculations show that electricity is generated at a cost of 3 to 12 US Cents/kW [9], which is quite high compared to electricity generated from other resources like hydel and fossil fuels. Thus, strong government initiatives are required to make solar power plants an attractive and viable option for electricity generation in the future.

8. Advantages of Solar Thermal in Pakistan

Solar thermal technology has got some remarkable advantages. This technology does not require resources other than solar energy. It is not dependent on resources which are going to deplete unlike conventional systems which generate electricity using non-renewable fossil fuels. These fossil fuels are rapidly depleting thus, there is an increasing need for alternate resources to generate power.

Environmental pollution has also become a major concern in recent times which requires regulation of greenhouse gases. Burning fossil fuels leads to emission of greenhouse gases (GHG). These gases are quite harmful for the environment and people [43]. The emission of gases due to burning of fossil fuels is one of the main causes for many diseases. These diseases include neurological problems, heart issues, cancer and diseases associated with breathing. GHG emissions are quite low in the case of solar energy when compared to coal and natural gas. CO_2 emission are in the range of 0.64 to 1.63 kg/kWh of energy generated from coal, 0.27 to 0.91 kg/kWh of energy generated from natural gas and 0.03 to 0.09 kg/kWh of energy generated from solar resources. These statistics show that the ratio of carbon emissions is 18:9.5:1. This clearly implies that solar energy is cleaner than coal and gas. Thus, power generation from solar resources can lessen the chances of these diseases. Some studies reveal that the use of renewable energy like solar energy instead of fossil fuels can decrease the number of wasted workdays, premature morality rates and health related expenditures [44].

Solar technologies also require more labour. This suggests that they can provide greater job opportunities to the people living in nearby communities. According to some reports, the solar industry in USA has provided up to 208,859 jobs to different people in different capacities [44]. Similarly, the implementation of solar technologies in Pakistan will definitely provide greater job opportunities to the people living in the country.

9. Limitations of Solar Thermal Technologies in Pakistan

There are also many limitations which are currently associated with the implementation of solar energy technologies in Pakistan. Solar energy systems need high initial capital cost which makes them unaffordable for many people across the country. They are also more expensive compared to other conventional electricity production systems like fossil fuel and nuclear energy based generations. Other challenges include manufacturing techniques and waste materials. Pakistan is a country with low literacy rates, due to which proper guidance for consumers about the opportunities and usage of solar energy appliances proves to be a hard and costly task. False hopes given to rural communities about electrifying rural areas by politicians also cause hurdles in the development of solar energy. It is also difficult to get proper data regarding environmental conditions, which is crucial for harnessing solar energy.

Besides other issues, the intermittent nature of the solar energy is also of prime concern. Solar energy is not considered to be a reliable source of energy because of its sporadic behaviour and thus is not a suitable choice for continuous power supply. Preferably, solar energy should be used in combination with another sort of power generation system to provide continuous power. If another power generation system is not used in combination with solar energy production, a proper storage system should be employed to ensure continuous power supply. Quality issues are also associated with solar energy.

Certain damages to environment are also possible due to poor handling of solar thermal power plants while in operation. Solar power tower technology can be dangerous for birds, insects and other animals flying in the course of concentrated solar rays. The high temperatures associated with the concentrated rays can kill these animals. The working fluids used in solar concentrators are sometimes harmful for humans and should be properly dealt with [1].

Solar thermal systems require the coolant to be changed over a specific period of time which poses a real challenge because it can be harmful for the surroundings. Apart from this, solar thermal systems require a large amount of dedicated land. Another issue related with these systems is they require large amount of water for cycle cooling, which is only possible when surrounding areas have an abundance of water. It has also been seen that solar energy potential is often more in deserts where there is scarcity of water, thus making it difficult to harness solar energy for thermal generation because water is frequently needed for thermal cycles and to clean surfaces of reflectors [15].

The efficiencies of solar thermal systems are quite low as these systems have not yet been properly developed. Based on the different power systems used, their efficiencies can range from 10 to 17%. Their initial cost is also very high which makes them undesirable for end customers. These systems work only during the day when solar energy is available. However, in most cases, peak loads occur at night time. Hence, a storage system should be available to provide energy during peak hours. Expensive batteries, which need to be changed every 3-5 years, are used for this purpose. These batteries account for 30 to 40% of the total cost. Furthermore, solar energy is highly dependent upon geographical location and plants cannot be installed unless detailed feasibility studies are carried out [45].

As solar energy is a form of renewable energy, it faces a lot of challenges. There are certain issues involved in the switch from conventional energy resources to renewable energy resources. Public awareness is one of the toughest hurdles when discussing the implementation of a renewable energy program. To counter this, multiple organizations and agencies should be brought in to action to make familiarize the public with renewable energy. Training and guidance programs should be also arranged on timely basis to spread the importance of the need for sustainable energy. Besides general public awareness, research and development activities are necessary. Universities can play a very crucial role in the development of sustainable energy. Research is quite important yet challenging due to the lack facilities. Regardless, various subjects regarding sustainable energy should be included in the curriculum. As of now, Pakistan does not have the proper infrastructure for renewable energy. Thus, proper infrastructure needs to be developed to commercialize renewable energy in the market. Financial incentives, if properly given can change the current power scenario of the country. Strict monitoring for the implementation of this program is also lacking in the country [46].

10. Institutions working for Development of Solar Energy in Pakistan

Solar energy can act as an ideal resource for power generation in the country since it can mitigate the power crises in the country. Many institutions have been made to monitor all activity regarding solar energy. These institutions are busy in research and development to promote solar energy in the country. Some of them also provide technical and financial assistance to promote solar energy. Some of these institutes are discussed below.

IOP Conf. Series: Materials Science and Engineering **899** (2020) 012008 doi:10.1088/1757-899X/899/1/012008

10.1 Pakistan Council of Renewable Energy Technologies (PCRET)

PCRET was brought into existence in 2001 by merging the National Institute of Silicon Technology (NIST) and the Pakistan Council for Appropriate Technologies (PCAT). The merging of these two institutes was necessary to carry out research and development activities in a better manner and to avoid any sort of complications regarding duplication of research activities. PCRET has established its head office in Islamabad and four regional offices in Quetta, Karachi, Lahore and Peshawar.

Currently, PCRET is of the most important institutions busy in promoting renewable energy in Pakistan. It is responsible for research and development of different renewable energy programs. PCRET is particularly active in promoting solar energy. Apart from conducting research activities related to PV modules, it has also done a remarkable job to promote solar thermal technology. PCRET has designed and fabricated many solar thermal based appliances. These appliances include solar water stills, solar cookers, solar dryers for drying of agricultural products, solar heaters for heating water and solar heating systems for space heating. These appliances are used for different domestic, commercial and industrial purposes across the country. Being a responsible organization, PCRET has worked and is currently working on different type of solar appliances to modify and improve them to achieve better results. Apricots are produced abundantly in the country and their preservation and dehydration is a major task. Other achievements of PCRET include design and fabrication of solar cookers and solar water stills. PCRET provided more than 500 solar cookers to non-government organizations (NGOs) for distribution among local communities to familiarize them with this technology. 50 solar water stills were also given to NGOs to provide them to residents of the country. This increases awareness regarding renewable energy which makes people less hesitant to switch to these systems in the future. Another prominent achievement of PCRET is the development and fabrication of a solar space heating system in the local market. This system provides heat to the PCRET building.

Solar dryer technology is used worldwide. Pakistan is also taking a keen interest to develop this technology. For this purpose, PCRET is playing a key role to develop and promote solar dryer technology across different regions of the country. A hybrid solar dryer has been designed and fabricated by PCRET to dehydrate apricots, which is a great achievement. Under another project, PCRET has fabricated and manufactured 10 solar dryers best suitable for communities in remote areas. These remote areas use these solar dryers for drying of dates. These dryers have capacity to store 500 kg of dates. Regions where these dryers have been installed include Jhang, Multan, Muzaffargarh and D.G. Khan districts of Punjab, Sukkur and Khairpur districts of Sindh, Turbat and Panjgoor districts of Baluchistan and Dera Ismail Khan district of KPK. Fabrication of these solar dryers has been done locally. This project was possible due to local community involvement. The cost of the project was also shared. The local community chose two persons who were fit for the job and they remained in touch with the project from starting till finishing to acquire the necessary training. These dryers dry an appreciable amount of 25 tons of dates each year. These dates in turn provide these communities with a 3.5 million revenue. These dryers are operated by local communities. Routine maintenance is also done by the communities. The technical assistance team of PCRET closely monitors and checks performance of these dryers to resolve any complications [9].

10.2 Solar Energy Research Center

Solar Energy Research Center is working under the Ministry of Science and Technology. It is situated in Hyderabad city. This center is mainly concerned with research activities in the field of solar energy. The research of this center is particularly focused on solar thermal applications. These solar thermal applications include solar desalination systems, solar heating and air conditioning and power generation from solar thermal technologies [8].

10.3 Alternate Energy Development Board (AEDB)
AEDB was established in 2003 under the supervision of the federal government. The prime objective of the AEDB was to enhance the development of renewable energy in the country. Another goal of the AEDB is to switch 5% power generation of the country to renewable and alternate energy by 2030. It was also responsible for electricity supply to 80,000 remote villages of the country through renewable and alternate resources of energy [32].

10.4 Renewable and Alternative Energy Association of Pakistan (REAP)
REAP is an institution currently working in different areas to improve and develop renewable and alternate energy in the country. It is specifically responsible for promoting renewable and alternate energies to decrease reliance on conventional fossil fuels for power generation [32].

10.5 EME College, NUST Islamabad
EME College is a part of NUST (National University of Sciences and Technology), located in the capital of the country, Islamabad. EME College is engaged in research and development activities regarding solar energy. This college has worked to install a solar water heater with a capacity of 40 tons/day for a local industry at Lahore. This college is also busy in activities regarding solar pumps. It is also working to electrify houses through solar energy. FATA has benefited from this facility. Work has been started to electrify 450 houses in FATA since May 2009. Other activities of this college include wind/solar hybrid water pumping systems. It also intends to install 50 such units in Kharan which is a district situated in Baluchistan. It also works to promote solar cookers and solar dryers. For this purpose, it has provided Bagh, Azad Kashmir with twelve solar cookers and two dryers, each with a 50 kg capacity [9].

10.6 Academic Institutions
Research activities are not focused on in a majority of the universities in Pakistan because research environment in Pakistan is still not developed. Same applies to research activities in the field of renewable energy, particularly solar energy. Regardless, some universities are busy in conducting research and development activities in the broad field of solar thermal energy. Some institutions are worth mentioning regarding research in solar thermal applications. Institute of Environmental Studies at Ghulam Ishaq Khan (GIK) Institute of Engineering Sciences and Technology in Topi, Swabi is currently engaged in the testing of solar thermal devices. Department of Mechanical engineering at University of engineering and technology (UET) Lahore is also conducting various research activities in the field of solar thermal technologies [8].

10.7 U.S.-Pakistan Center for Advanced Studies in Energy (USPCAS-E), NUST Islamabad
This center has been established at NUST Islamabad. It is a joint venture by NUST and USAID. This center is focused on the development of research and development activities in the country, with the help of USAID, to make country secure in terms of energy. The primary focus areas of research in this center are renewable and emerging technologies [32].

10.8 U.S.-Pakistan Center for Advanced Studies in Energy (USPCAS-E), University of Engineering and Technology (UET), Peshawar
UET Peshawar has established this center in collaboration with USAID to conduct research and development activities in the field of energy. Despite being rich in natural resources, the country is experiencing a power deficit. This center is busy in conducting different research activities in the field of energy. For this purpose, it has started several projects in different areas of the country. It has also initiated several projects with local communities. Local communities are involved in sharing responsibilities which familiarizes them with renewable energy. This center is also busy in educating students through different

postgraduate studies programs. Renewable Energy Engineering is one of these programs which is purely based and focused on renewable energy.

11. Conclusion

Pakistan is facing a severe energy crisis at the moment. There are many regions across the country where load shedding has become a routine matter. Due to routine power cutoffs, the residents of the country, especially those residing in rural areas, are severely affected since electricity dependent domestic chores cannot be executed properly. Businesses and commercial activities are also disturbed due to uninformed load shedding schedule.

In these circumstances, the country is in a dire need for power generation. Currently, the country relies on expensive, non-renewable fossil fuels for power generation, which are depleting rapidly. Moreover, they are quite hazardous for the environment since they produce greenhouse gases upon burning. These greenhouse gases pollute the environment and cause multiple diseases. Considering these circumstances, power generation in the country should shift to renewable energy resources.

Solar energy is one of the main sources which can be used to mitigate the power crises in the country. Pakistan is blessed with a tremendous amount of solar energy potential which needs to be harnessed. Solar energy is clean and does not pollute the atmosphere upon usage. Pakistan has got multiple regions with sufficient solar intensity for power generation through solar thermal power plants. Furthermore, the country has a large area available for the installation of these plants. These will not only produce power, but will also help in controlling greenhouses gases emissions. At the moment, the cost of solar thermal power plants is higher compared to conventional power plants. However, solar thermal power plants can provide affordable electricity, if the government provides sufficient subsidies similar to those provided to conventional power plants.

Most of the people living in remote areas of the country are poor. Solar thermal appliances can prove to be most beneficial to these groups of people. Solar cooking will help these people, not only economically, but also by ridding them of the need to collect fuels like wood, which will improve the standard of living in these rural communities. The solar pump can be a vital technology for remote areas which are not connected to the electrical grid and lack water. This technology will provide ease of access to water, leading to social improvement. Solar drying is also a useful technique which can help people in remote areas in the preservation fruits and vegetables, which in turn can act as great source of income for these communities. Besides that, agricultural products will not be wasted due to rotting, which will help in reducing the scarcity of food in the country. The advantages of this technology can be availed only if the residents of the country are properly trained and familiarized with them. Solar thermal technologies can be amongst the best options for the country in the near future, if proper attention is given.

References

[1] Devabhaktuni V, Alam M, Depuru S S S R, Green II R C, Nims D, Near C J R and Reviews S E 2013 Solar energy: Trends and enabling technologies **19** 555-64

[2] Jabeen M, Umar M, Zahid M, Rehaman M U, Batool R, Zaman K J R and Reviews S E 2014 Socio-economic prospects of solar technology utilization in Abbottabad, Pakistan **39** 1164-72

[3] Khan M A, Latif N J R and Reviews S E 2010 Environmental friendly solar energy in Pakistan's scenario **14** 2179-81

[4] Muneer T, Maubleu S, Asif M J R and Reviews S E 2006 Prospects of solar water heating for textile industry in Pakistan **10** 1-23

[5] Schiermeier Q, Tollefson J, Scully T, Witze A and Morton O J N N 2008 Energy alternatives: Electricity without carbon **454** 816-23

[6] Asif M 2009 Sustainable energy options for Pakistan Renewable and Sustainable Energy Reviews. Volume)

[7] Tian Y and Zhao C-Y J A e 2013 A review of solar collectors and thermal energy storage in solar thermal applications **104** 538-53
[8] Mirza U K, Maroto-Valer M M, Ahmad N J R and Reviews S E 2003 Status and outlook of solar energy use in Pakistan **7** 501-14
[9] Bhutto A W, Bazmi A A, Zahedi G J R and Reviews S E 2012 Greener energy: issues and challenges for Pakistan—solar energy prospective **16** 2762-80
[10] Ellabban O, Abu-Rub H, Blaabjerg F J R and Reviews S E 2014 Renewable energy resources: Current status, future prospects and their enabling technology **39** 748-64
[11] Shukla R, Sumathy K, Erickson P, Gong J J R and Reviews S E 2013 Recent advances in the solar water heating systems: A review **19** 173-90
[12] Farzaneh A, Mohammadi M, Ahmad Z and Ahmad I 2013 Aluminium Alloys in Solar Power–Benefits and Limitations
[13] Garud S, Purohit I J T E and Resources Institute D S B, IHC Complex, Lodhi Road, New Delhi 2009 Making solar thermal power generation in India a reality–overview of technologies, opportunities and challenges **110003**
[14] Busse K T and Dinter F J J o E i S A 2016 Overview of predictive CSP spread prospects and its opportunities **27** 50-9
[15] Machinda G, Chowdhury S, Arscott R, Chowdhury S and Kibaara S 2011 Concentrating solar thermal power technologies: a review. In: *2011 Annual IEEE India Conference*: IEEE) pp 1-6
[16] Farooqui S Z J R and Reviews S E 2014 Prospects of renewables penetration in the energy mix of Pakistan **29** 693-700
[17] Raja N K, Khalil M S, Masood S A and Shaheen M J J o A S 2011 Design and manufacturing of parabolic trough solar collector system for a developing country Pakistan **7**
[18] Dabiri S and Rahimi M F Basic introduction of solar collectors and energy and exergy analysis of a heliostat plant
[19] Kesari J, Gupta M, Jain A and Ojha A J I J R S I 2015 Review of the concentrated solar thermal technologies: challenges and opportunities in India **2** 105-11
[20] Srilakshmi G, Venkatesh V, Thirumalai N, Suresh N J R and reviews s e 2015 Challenges and opportunities for Solar Tower technology in India **45** 698-709
[21] Kalogirou S J R P o E 2007 Recent patents in solar energy collectors and applications **1** 23-33
[22] Bodhle G 2016 Solar Thermal Power Technology
[23] Solar plants make a significant contribution to sustainable, climate-friendly energy generation. schlaich bergermann partner)
[24] Poullikkas A, Hadjipaschalis I, Kourtis G J T i H and Transfer M 2013 A comparative overview of wet and dry cooling systems for Rankine cycle based CSP plants **13** 27
[25] Aboelwafa O, Fateen S-E K, Soliman A, Ismail I M J R and Reviews S E 2018 A review on solar Rankine cycles: Working fluids, applications, and cycle modifications **82** 868-85
[26] Desai N B, Kedare S B and Bandyopadhyay S J S E 2014 Optimization of design radiation for concentrating solar thermal power plants without storage **107** 98-112
[27] Farooq M, Shakoor A J J o R and Energy S 2013 Severe energy crises and solar thermal energy as a viable option for Pakistan **5** 013104
[28] Awan A B, Khan Z A J R and Reviews S E 2014 Recent progress in renewable energy–Remedy of energy crisis in Pakistan **33** 236-53
[29] Löf G O J G t t R o t S E A L and html S V F a h l c e a f u u 1963 Recent investigations in the use of solar energy for cooking
[30] Muthusivagami R, Velraj R, Sethumadhavan R J R and Reviews S E 2010 Solar cookers with and without thermal storage—a review **14** 691-701
[31] Belessiotis V and Delyannis E J S e 2011 Solar drying **85** 1665-91

[32] Bakhtiar F, Ahmed A J S, Technology, and Development 2017 A review of solar energy in Pakistan: Current status and future prospects **36** 189-95

[33] WEC W E C 2000 Renewable energy in South Asia: Status and prospects. World Energy Council (WEC) London)

[34] Fudholi A, Sopian K, Ruslan M H, Alghoul M, Sulaiman M J R and reviews s e 2010 Review of solar dryers for agricultural and marine products **14** 1-30

[35] Sukhera M J S e 1984 Utilization of solar energy—a programme for the development of Cholistan desert **33** 233-5

[36] Samee M A, Mirza U K, Majeed T, Ahmad N J R and Reviews S E 2007 Design and performance of a simple single basin solar still **11** 543-9

[37] Arjunan T, Aybar H, Nedunchezhian N J R and Reviews S E 2009 Status of solar desalination in India **13** 2408-18

[38] Liu L-q, Wang Z-x, Zhang H-q, Xue Y-c J R and Reviews S E 2010 Solar energy development in China—A review **14** 301-11

[39] Velmurugan V, Srithar K J R and reviews s e 2008 Prospects and scopes of solar pond: a detailed review **12** 2253-63

[40] Ullah I, Rasul M G, Sohail A, Islam M and Ibrar M J T P P-A A 2013 Feasibility of a Solar Thermal Power Plant in Pakistan

[41] Shinwari W K, Ali F and Nayyar A J T P D R 2004 Electric power generation from solar photovoltaic technology: is it marketable in Pakistan? 267-94

[42] De Graaff P J 1991 A multicylinder free-piston Stirling engine for application in Stirling-electric drive systems. In: *IECEC'91; Proceedings of the 26th Intersociety Energy Conversion Engineering Conference, Volume 5,* pp 205-10

[43] Khan J, Arsalan M H J R and Reviews S E 2016 Solar power technologies for sustainable electricity generation–A review **55** 414-25

[44] Kabir E, Kumar P, Kumar S, Adelodun A A, Kim K-H J R and Reviews S E 2018 Solar energy: Potential and future prospects **82** 894-900

[45] Dwivedi A, Bari A, Dwivedi G J I J o E R and Technology 2013 Scope and Application of Solar Thermal Energy in India-A Review **6** 315-22

[46] Aziz M F and Abdulaziz N 2010 Prospects and challenges of renewable energy in Pakistan. In: *2010 IEEE International Energy Conference*: IEEE) pp 161-5

IOP Conf. Series: Materials Science and Engineering **899** (2020) 012009 doi:10.1088/1757-899X/899/1/012009

Data based sensing of Shale Oil yield in Oil Shale Retorting process

Hasan Qayyum Chohan[1] and **Iftikhar Ahmad**[1,*]

[1]School of Chemical and Materials Engineering, National University of Sciences and Technology, Islamabad, Pakistan

*Email: iftikhar.salarzai@scme.nust.edu.pk

Abstract. Oil shale is sedimentary organic rocks that are being converted into useful shale oil and shale gas. North American regions, Canada and China are exploring the oil shale reserves to accommodate the depletion of natural oil and gas resources. Oil shale retorting technology is being utilized to convert the shale rocks into shale oil and shale gas. The major product is oil that is further treated to convert it into gaseous form. In this study, machine learning techniques like ensemble learning (least square boosting and bagging) and artificial neural network (ANN) are employed for data sensing of oil shale retorting process and being compared. Data is generated for ensemble models through MATLAB-Excel-Aspen interfacing. The proposed framework shows that ANN provides higher accuracy as compare to other models for oil shale retorting process for efficient oil recovery.
Keywords: Oil Shale, Retorting; Machine learning; Artificial neural network; Ensemble learning; MATLAB; Excel; Aspen.

1. Introduction

Energy is an integral part of world. Energy resources are depleting day by day due to significant advancements and developments in economics and population growth. Therefore, world is focusing to resolve the energy challenges and looking for alternative solutions. Oil and gas reservoirs / petroleum reservoirs are based on hydrocarbons present in porous and fractured rocks. They are classified into conventional as well as unconventional petroleum reservoirs. Conventional reservoirs of oil and gas are trapped in low permeable rocks while unconventional rocks have high porosity factor. The naturally occurring hydrocarbons are trapped in rocks due to high porosity factor. Extraction of oil and gas requires extensive exploration methods for efficient recovery of oil and gas present in underlying rocks. Pyrolysis of organic component 'kerogen' present in sedimentary rocks converts it into oil and gas. The sedimentary rock is called oil shale and product is known as 'shale oil' and 'shale gas' [1, 2]. In-situ and ex-situ techniques are employed to convert the oil shale into desired shale oil and other industrial fuels.

According to worldwide production reports of oil and gas 2018 presented in oil and gas journal (OGJ), 165.1 trillion bbl and 69,550 tcf oil and gas are produced per year. World is dependent on naturally occurring oil resources. Canada, USA and China are the key players in exploration of oil and gas through 'oil shale' [3]. Institute of energy research provides an overview of countries having major oil shale resources shown in Figure 1 [4].

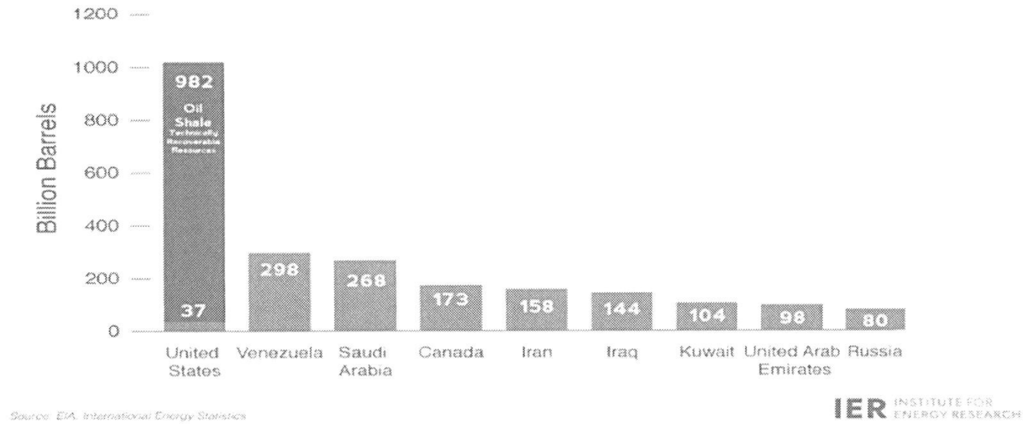

Figure 1. Oil Shale Reserves Worldwide [4]

Modeling and simulation is playing an important role in the industrial processes for optimization and reliability. Oil shale has become an interesting research area in many countries. In literature, various studies have been reported. Luo et al. examined the performance of oil shale processing of hydrogenation technology and analyzed the process of hydrogen through exergy analysis and found 69.20% exergy efficiency of simulated model [5]. Zhou et al. addressed the problems related to low utilization rate of oil shale raw material, low yield and heavy components presence. Techno- model economic was performed with addition of an indirectly heating moving bed in simulation model. The simulation modeling resulted a decrease of 12% in cost of oil shale production as compared to conventional processing [6]. Mu et al. studied parametric study of oil shale retorting for exergy analysis on China comprehensive utilization system [7]. Wang et al. determined the environmental sustainability and resource utilization efficiency factors through energetic life cycle assessment by Aspen Plus of Fushun type oil shale retorting. [8].

Yang et al. utilized the MATLAB interfacing with Aspen Plus to evaluate technical and economic aspects of solid heat carrier technology in oil shale retorting process [9]. Raja et al. investigated the presence of high valued hydrocarbon fuels and hydrogen in oil shale retorting process using Aspen Plus V8.4 modeling. [10]. Li et al. compare two technologies of oil shale retorting Dagong and Fushun. Exergy analysis through Aspen Plus [11]. Tamm et al. also observed thermodynamic parameters of three different samples of oil shale using Aspen Plus V8.6 [12]. Lv et al. studied the conversion of oil shale into bio fuel. The exergy analysis was performed using Aspen Plus [13]. Li et al. studied the effect of deposition of carbon on the overall performance of oil shale retorting process [14].

In this study, soft sensor is developed using machine learning techniques like ensemble learning and artificial neural network for oil shale retorting technology. Data based soft sensor is developed to predict the product composition as well as yield of shale oil. Aspen Plus V10.0 based model is used to generate the data using interface of Aspen Plus MS Excel 2010 and MATLAB R2018a. This work predicts our quantities of interest i.e. oil yield through retorting process and compare different machine learning techniques.

Section II of this study explains the process description for oil shale retorting process followed by modeling techniques in section III. Section IV describes the results and discussion while section V concludes the work.

2. Process description

Oil Shale is converted into shale oil (constituent of crude oil) through retorting process. Ammer et al. developed Aspen Plus V10.0 model for retorting as given in Figure 2 [15]. The major component of oil shale 'Kerogen' is crushed into small pieces to increase surface area of particles for efficient reaction.

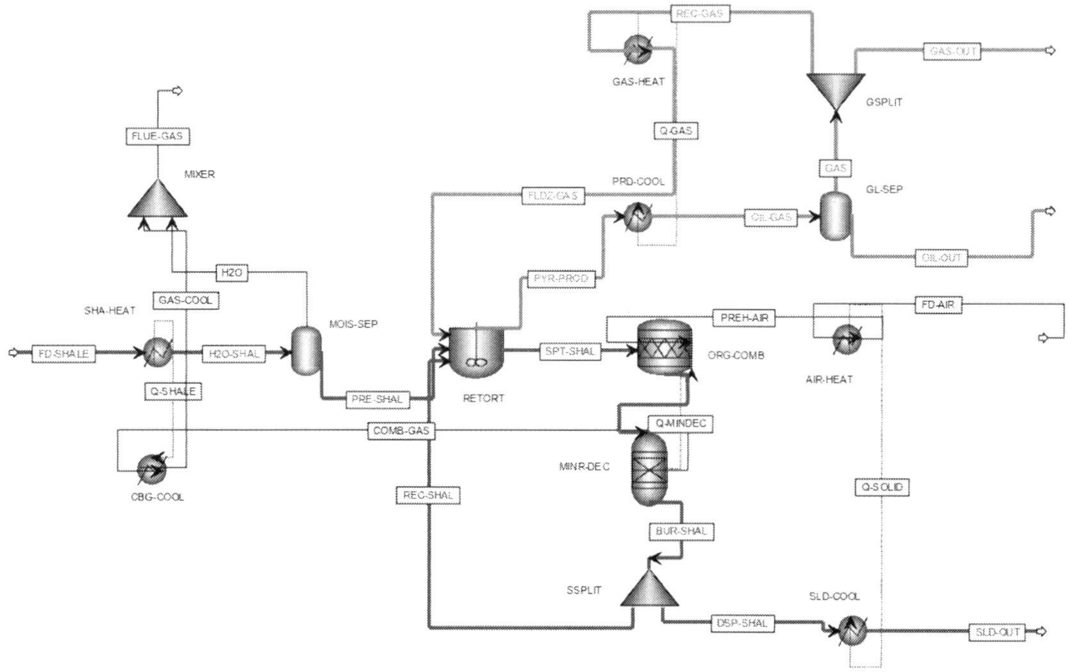

Figure 2. Aspen Plus Model of Oil Shale Retorting Process [15]

Oil Shale (FD-SHALE) is heated with in a pre-heater (SHA-HEAT) and moisture is removed through (MOIS-Sep). Pre-heated oil shale is entered in fluidized bed equipment i.e. called "Retort". Oil shale is converted to gas, oil and char in retort by providing temperature up to 950°F. The temperature requirement is accomplished with help of burned shale in combustor (ORG-COMB) using hot combustible gas (COMB-GAS) coming from MINR-DEC. Air (FD-AIR) is introduced to through heater (AIR-HEAT) for combustion reactions. The spent shale is expelled to a combustor from retort in which hot air coming from pre-heater decomposes spent shale minerals like dolomite and calcite etc. The hot gas of combustor is used for pre-heating of oil shale while some part of burned shale is recycled to the retort through splitter (SSPLIT) and other is separated for disposal (SLD-OUT). The product gas is sent to fractionator (GL-SEP) to recover the final product shale oil (OIL-OUT). The other components are recycled to retort and further processed for ultimate oil recovery. Recycle streams are also utilized for heat transfer mechanism to increase the exergy of process.

3. Machine learning techniques

3.1. Ensemble learning
Ensemble learning models are developed using least square boosting and bagging techniques. Boosting and bagging develop efficient models using combination of different weak models. The concept of ensemble learning models is demonstrated in Figure 3 [16]. The models are trained on stochastic data in several rounds and then models predict data by increasing their respective aggregated weights.

3.2. Artificial neural network
Artificial neural network (ANN) is biological brain type neural network that develops machine learning models to predict the accuracy of stochastic data. Neurons of network receive inputs and predict outputs using transfer function as given in Figure 4 [17]. The accuracy of predicted model is achieved by optimizing number of hidden layers. The increase of hidden layers in network and presence of more stochastic data provides more accuracy in predicted data.

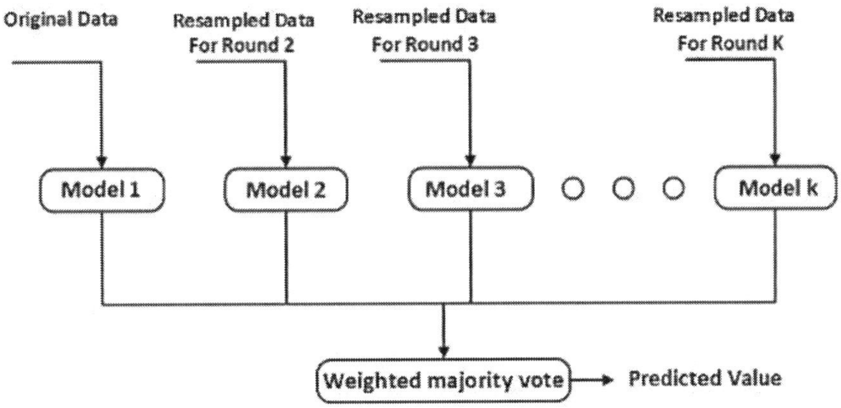

Figure 3. Ensemble Learning Framework [16]

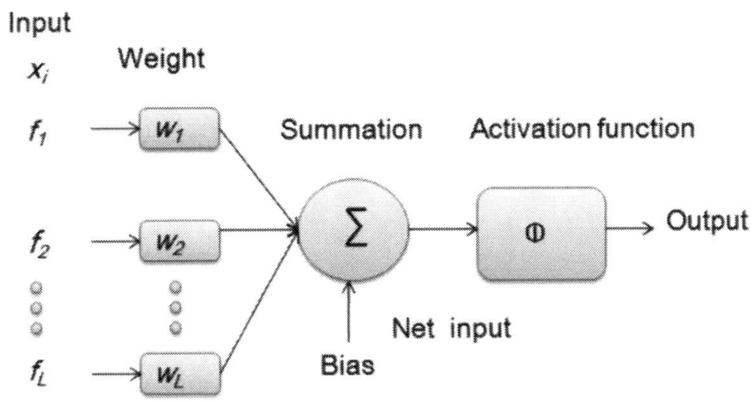

Figure 4. ANN Framework [17]

4. Results and discussions

This section deals with the results obtained for production of shale oil yield through retorting process. The results are also provided details of carbon dioxide extracted in flue gases. The results are derived using machine learning techniques i.e. least square boosting, bagging and artificial neural network. The model is developed using 13 input variables given in Table 1. Oil shale and air temperature, elemental composition of kerogen, retort and fractionator specifications are varied in Aspen Plus model to predict the machine learning models. The composition of illite is also altered to determine the impact on production of oil through retorting. Shale oil flow rate and carbon dioxide flow rate in flue gases are output variables enlisted in Table 2.

Table 1. Process input variables

No.	Process variable	No.	Process variable
1	Air flowrate	11	GL-Separator Temperature
2	Illite mass fraction	12	GL-Separator Pressure
3	Silica mass fraction	13	G-Split Fraction
4	Carbon Composition		
5	Hydrogen Composition		
6	Shale Heater Temperature		
7	Air Heater Temperature		
8	Retort Pressure		
9	Retort Volume		
10	Retort vol. fraction		

Table 2. Process output variables

No.	Process Variable
1	Oil flowrate (Product)
2	CO_2 flow rate (Flue Gases)

Ensemble models are developed for oil recovery and carbon dioxide in flue gases. The results are being compared for three different ensemble learning techniques. Data is generated for 1350 cases using MATLAB-Excel-Aspen interfacing in which input variables are varied at some extent. Data is divided into training and testing data. Training data is utilized for model development while 20% testing data to validate ensemble model. The modeling framework is given in Figure 5.

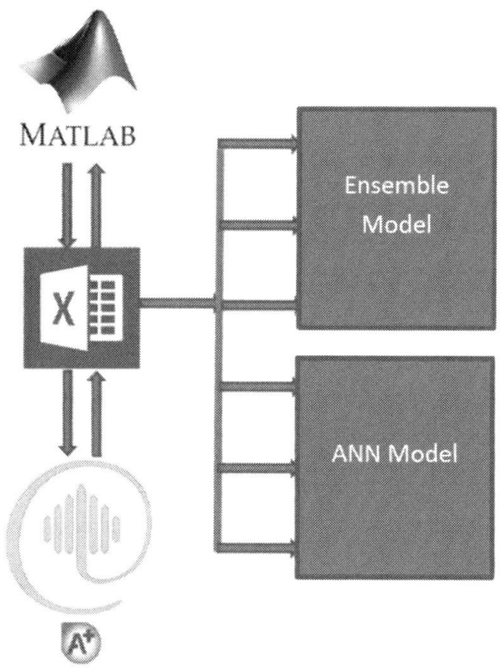

Figure 5. Modeling Framework

Data sensing accuracy in prediction of shale oil in final product is 97.8%, 99.8% and 99.6% respectively for boosting, bagging and ANN models. Similarly, for carbon dioxide, boosting, bagging and ANN models provide accuracy up to 98.9%, 99.7% and 99.9% respectively. The results for three ensemble models are given in Figure 6-11. Predicted data is optimized by increasing or decreasing of ensemble layers. For better accuracy of artificial neural network predicted data, hidden layers of neurons are increased. The presence of maximum stochastic data of industrial operations helps more in data sensing and much accurate prediction.

Root mean square error values also exhibit that bagging model is given much accuracy with lowest value of RMSE for oil production (0.0597) as compare to LSBoost (0.2467) and ANN (0.1033). Table 3 shows the exhibiting values of accuracy and RMSE for three models. Table 4 shows the accuracy of validated data of oil shale retorting process.

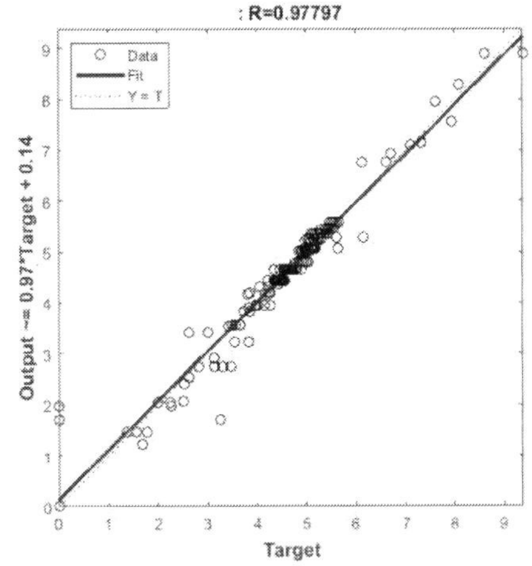

Figure 6. LS Boosting Regression Performance of Oil

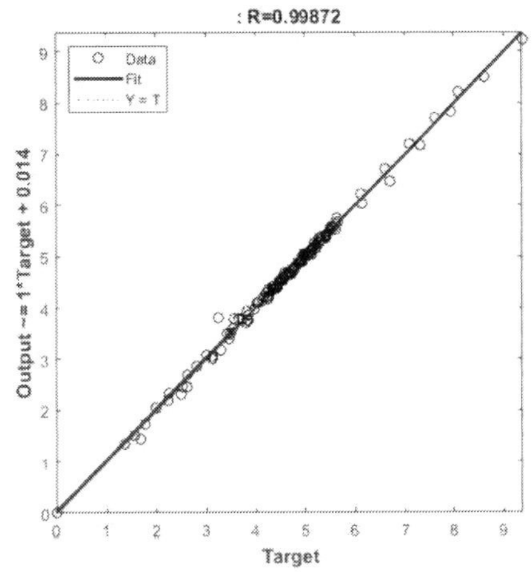

Figure 7. Bagging Regression Performance of Oil

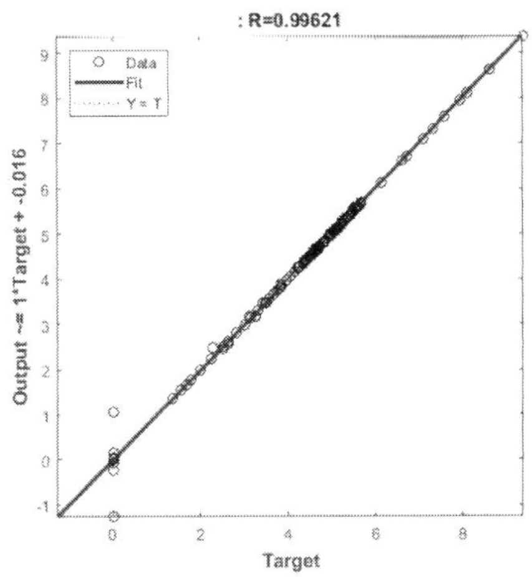

Figure 8. ANN Regression Performance of Oil

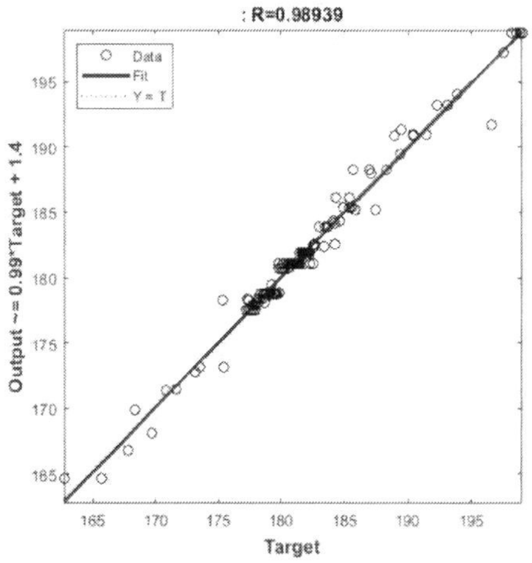

Figure 9. LS Boosting Regression Performance of CO_2

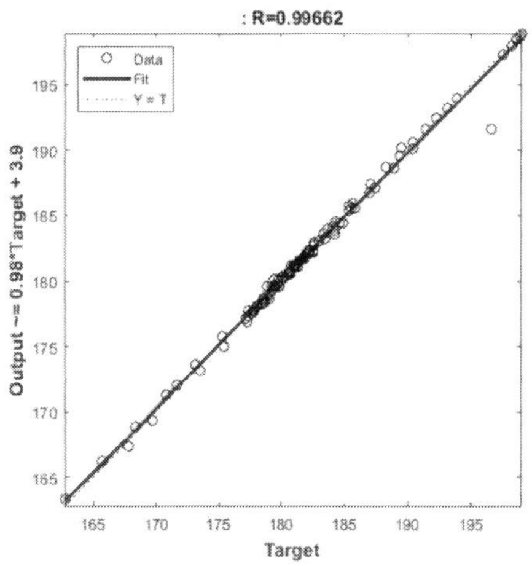

Figure 10. Bagging Regression Performance of CO_2

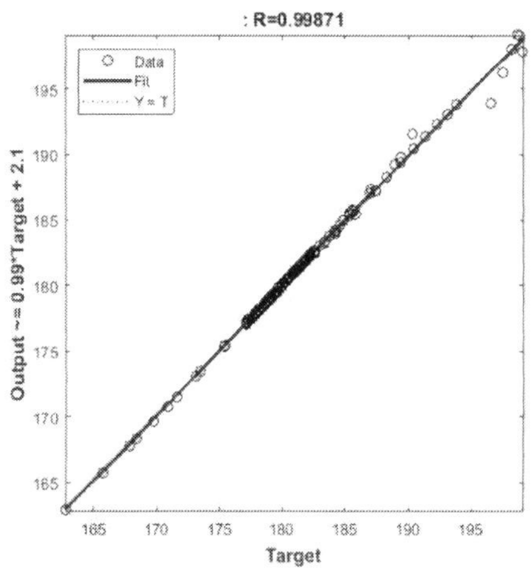

Figure 11. ANN Regression Performance of CO_2

Table 3. RMSE values for output variables

No.	Process variable	LSBoost	Bag	ANN
1	Oil flowrate (Product)	2.467e-1	5.97e-2	1.03e-1
2	CO_2 flow rate (Flue Gases)	6.143e-1	3.509e-1	2.169e-1

Table 4. Regression values for output variables

No.	Process variable	LSBoost	Bag	ANN
1	Oil flowrate (Product)	97.8%	99.8%	99.6%
2	CO_2 flow rate (Flue Gases)	98.9%	99.7%	99.8%

5. Conclusions

In this study, ensemble learning models of least square boosting and bagging are compared with artificial neural network model. The results predicted that ANN provides more accuracy as compare to least square boosting while bagging is providing comparable behavior to ANN. The predicted data show that machine learning techniques are efficient methods to evaluate the accuracy for real time applications in industries.

6. Nomenclature

ANN	Artificial Neural Network
OGJ	Oil and Gas Journal
FD-SHAL	Oil Shale Feed
SHA-HEAT	Shale Pre-Heater
MOIS-Sep	Moisture Separator
ORG-COMB	Combustion Chamber
COMB-GAS	Combustible Gas
FD-AIR	Air Inlet Stream
AIR-HEAT	Air Pre-Heater
SSPLIT	Burned-Shale Splitter
SLD-OUT	Spent Shale
GL-SEP	Gas-Oil Fractionator
OIL-OUT	Product (Oil)

7. References

[1] Zhi Y, Caineng Z, Songtao W, Senhu L, Songqi P, Xiaobing N, Guangtian M, Zhenxing T, Guohui L and Jiahong Z 2019 Formation, distribution and resource potential of the" sweet areas (sections)" of continental shale oil in China Marine and Petroleum Geology 102 48-60
[2] Seewald J S 2003 Organic–inorganic interactions in petroleum-producing sedimentary basins Nature 426 327
[3] Guo H, Dong J, Wang Z, Liu H, Ma R, Kong D, Wang F, Xin X, Li Y and She H 2018 2018 EOR Survey in China-Part 1. In: SPE Improved Oil Recovery Conference: Society of Petroleum Engineers)
[4] Golpour H and Smith J 2017 Oil Shale Ex-Situ Process-Leaching Study of Spent Shale International Journal of Engineering and Science Invention 6 45-53
[5] Luo X, Guo Q, Zhang D, Zhou H and Yang Q 2018 Simulation, exergy analysis and optimization of a shale oil hydrogenation process for clean fuels production Applied Thermal Engineering 140 102-11
[6] Zhou H, Zeng S, Yang S, Xu G and Qian Y 2018 Modeling and analysis of oil shale refinery process with the indirectly heated moving bed Carbon Resources Conversion

[7] Mu M, Han X and Jiang X 2018 Combined fluidized bed retorting and circulating fluidized bed combustion system of oil shale: 3. Exergy analysis Energy 151 930-9

[8] Wang Q, Ma Y, Li S, Hou J and Shi J 2018 Exergetic life cycle assessment of Fushun-type shale oil production process Energy Conversion and Management 164 508-17

[9] Yang Q, Li X, Qian Y and Zhang D 2018 Technical and economic analysis of an oil shale comprehensive utilization process with solid heat carrier technology Carbon Resources Conversion

[10] Raja M A, Chen H, Zhao Y, Zhang X and Zhang S 2017 Process simulation and assessment of hydrogen and high valued hydrocarbon fuels products from oil shale International Journal of Hydrogen Energy 42 4922-34

[11] Li X, Zhou H, Wang Y, Qian Y and Yang S 2015 Thermoeconomic analysis of oil shale retorting processes with gas or solid heat carrier Energy 87 605-14

[12] Tamm K, Kallaste P, Uibu M, Kallas J, Velts-Jaenes O and Kuusik R 2016 LEACHING THERMODYNAMICS AND KINETICS OF OIL SHALE WASTE KEY COMPONENTS Oil Shale 33

[13] Lv Q, Yue H, Xu Q, Zhang C and Zhang R 2018 Quantifying the exergetic performance of bio-fuel production process including fast pyrolysis and bio-oil hydrodeoxygenation Journal of Renewable and Sustainable Energy 10 043107

[14] Li H, Guo C and Yuan D 2017 Study on Carbon Deposition Characteristics of Oil Shale Dry Distillation Gas and Influencing Factors of Carbon Deposition Reaction Journal of Nanoelectronics and Optoelectronics 12 1339-42

[15] Ammer J 1986 Identification of data gaps found during the development of a zero-order model for a fluidized-bed retort/combustion process. USDOE Morgantown Energy Technology Center, WV)

[16] Ahmad I, Ayub A, Mohammad N and Kano M 2019 Data-Based Prediction and Stochastic Analysis of Entrained Flow Coal Gasification under Uncertainty Sensors 19 1626

[17] Ahmad I, Mabuchi H, Kano M, Hasebe S, Inoue Y and Uegaki H 2011 Data-Based Fault Diagnosis of Power Cable System: Comparative Study of k-NN, ANN, Random Forest, and CART IFAC Proceedings Volumes 44 12880-5

Novel Simplified Blade Hub Line Optimization of High-Pressure Ratio Centrifugal Compressor Design Using Numerical Simulations

A Khan[1*], F Ahmad[2], S Ali[1], M R Siddiqi[1] and M Z Ijaz[1]

[1]Department of Mechanical Engineering, CECOS University of IT and Emerging Sciences, Peshawar, Pakistan
[2]Department of Mechanical and Aerospace Engineering, Air University, Islamabad, Pakistan

[1*]E-mail: amjid@cecos.edu.pk

Abstract. The focus of this research is to numerically investigate the effect of blade hub line variation on the performance (Total pressure-ratio and Isentropic-Efficiency) of centrifugal compressor from stall to choke to study the operating range and stall margin. An optimization technique is carried out in which the meridional profile hub line is modified and compared with the high Mach number SRV2 compressor designed and fabricated by DLR (German Aerospace Center). Numerical simulations showed significant increase in the stall margin and operating range by bargaining on pressure ratio and isentropic efficiency. Reynolds Averaged Navier Stokes (RANS) based k-ε model is used to predict turbulence using numerical simulations. The value of Y plus for the structured mesh near the boundaries is kept 35. Blade hub line being the important parameter, has substantial performance improvement of centrifugal compressor. The novel design improved the stall margin by 44 percent while operating range from stall to choke has been upgraded by 7.5 percent.

1. Introduction

A mechanical device used to increase the total pressure of a ideal air fluid is called centrifugal compressor. Applications of centrifugal compression systems include gas turbine engines, turbochargers, powering tanks and rotorcraft, chemical industries, compression and transportation of hydrogen in petrochemical refineries, and refrigeration systems. Centrifugal compressors are used for low-flow rate applications because it has higher pressure ratio to weight ratio as compared to axial compression systems [1].

Optimization technique in the meridional profile is one of the important parameters in selection of centrifugal compressors for the specific applications [2]. When high pressure rise at low mass flow rate is desired, centrifugal compressor is considered for such applications [3]. For turbochargers, centrifugal compressors with high operating range and high stall margin is desired [4]. Hideo Nishida et al. investigated the blade loading optimization by increasing the back-sweep angle. This increase in the back-sweep angle is used to suppress the reduction in the surge margin and test results showed that the isentropic efficiency has been increased by 5% compared with the standard design [5]. For turbochargers applications, centrifugal compressors with high stall margin is required but at low mass flow rate flow instabilities leads to high aerodynamic loss and structural damage [6]. These flow instabilities are stall and surge [7]. The flow separation and flow recirculation due to adverse pressure gradient is called stall and surge is divided into two types i.e. mild surge and deep surge depending on the flow fluctuation and back flow condition [8].

Different optimization techniques have been carried out to improve the performance of the centrifugal compressor. Tip clearance optimization is carried out to observe its effect on the pressure ratio and efficiency of the compressor. At constant tip clearance of 0.1mm highest pressure ratio and efficiency is found [9]. Amjid et al. numerically investigated the influence of diffuser exit width on the performance of centrifugal compressor and it is found that increasing the diffuser exit decreases the pressure ratio and efficiency due to increase of entropy generation at the outlet of the impeller [10].

C. Xu and Amano studied the meridional impacts of the centrifugal compressor and they found that changing the meridional profile one way or another, it affects the performance of the compressor [11]. The study further shows that a proper meridional profile can improve overall performance of centrifugal compressor and reduce the sensitivity of the tip clearance [12]. T. Raitor et al. studied the geometry optimization of the impeller blade with 45 free parameters. The optimized design of the impeller improved the efficiency by 1.9 percent and operating range was also increased significantly. The forward sweep angle of the blade reduced the shock losses at the leading edge of the impeller [13]. Hong Xie et al. modified the meridional plane for the high operating range and efficient design and results showed that by changing the meridional profile of the compressor greatly effects the performance of the centrifugal compressor by increasing the stall margin and efficiency [2]. Amjid et al. numerically studied the effect of varying meridional profile and thickness and they concluded that increasing back-sweep angle at the leading edge increases the stall margin by 0.8%, optimizing the blade angle at the trailing edge increases the stall margin by 1.1% and increasing the thickness of the blade increases the stall margin by 0.2% [4].

1.1. Specifications of Test Case Compressor

A centrifugal compressor test case having high pressure ratio, mass flow coefficient and specific speed has been studied to understand effect of varying meridional profile, blade angles and thickness. The test case was designed and built by DLR (German Aerospace Center). The basic geometric parameters have been documented in Table 1.

As shown in the Fig. 2, impeller with13 full and 13 splitter blades. Splitter blades leading edges is at 26% of full blade chord. The exit diameter of test case compressor is 112 mm and nominal tip speed is 586 m/s at 50000 1/min. A vaneless diffuser is connected after the impeller [14].

The compressor (SRV2-O) test case data is shown in table:

Table I: Specifications of Test Case Compressor

Parameter	**Value**	**Units**
Inlet ambient pressure, P_{t1}	101325	[Pa]
Inlet ambient temperature, T_{t1}	288.15	[K]
Design Shaft speed, n	50000	[1/min]
Design mass flow rate, m	2.55	[kg/s]
Number of full/splitter blades	13/13	-
Tip speed of Impeller	586	[m/s]
Pressure ratio of Impeller	6.1	-
Impeller efficiency	84%	-

2. Numerical Setup

For CFD model to acquire, a step by step methodology has been taken on for numerical simulations. The model development consists of following steps such as Geometric parameterization in BladeGen, Computational method for meshing in TurboGrid, Computational solver, Turbulence model and boundary conditions in ANSYS CFX.

2.1. Governing Equations

The succeeding postulates based on the theoretical background are considered to shape a three-dimensional fluid flow model using a cartesian co-ordinate system [15]:

1) The presided flow is considered as continuum, and a no slip boundary condition is valid
2) The flow is compressible ideal flow
3) The flow is studied as turbulent because of high Reynolds number of $3.65x10^6$
4) A steady-state fluid flow is considered
5) The compressor walls are considered as completely smooth
6) The viscous dissipation effect is considered in the energy equations as it is very important in case of compressor to study viscous sublayer for flow separation
7) Compressor stage geometry is same as experimental setup
8) Turbulence model is K- ε model
9) Design mass flow rate 2.55kg/s
10) Design speed of 50,000 1/min

Based on postulates, a mathematical model is designed using the following equations [15]:

Continuity Equation:

$$\frac{\partial u}{\partial x}+\frac{\partial v}{\partial y}+\frac{\partial w}{\partial z}=0 \qquad (1)$$

Where u, v and w are the velocity components and x, y and z are cartesian co-ordinates.

Momentum Equation:

In the x-direction,

$$u\frac{\partial u}{\partial x}+v\frac{\partial u}{\partial y}+w\frac{\partial u}{\partial z}=-\frac{1}{\rho_f}\frac{\partial p}{\partial x}+\frac{\mu_f}{\rho_f}\left(\frac{\partial^2 u}{\partial x^2}+\frac{\partial^2 u}{\partial y^2}+\frac{\partial^2 u}{\partial z^2}\right) \qquad (2)$$

In the y-direction,

$$u\frac{\partial v}{\partial x}+v\frac{\partial v}{\partial y}+w\frac{\partial v}{\partial z}=-\frac{1}{\rho_f}\frac{\partial p}{\partial y}+\frac{\mu_f}{\rho_f}\left(\frac{\partial^2 v}{\partial x^2}+\frac{\partial^2 v}{\partial y^2}+\frac{\partial^2 v}{\partial z^2}\right) \qquad (3)$$

In the z-direction,

$$u\frac{\partial w}{\partial x}+v\frac{\partial w}{\partial y}+w\frac{\partial w}{\partial z}=-\frac{1}{\rho_f}\frac{\partial p}{\partial z}+\frac{\mu_f}{\rho_f}\left(\frac{\partial^2 w}{\partial x^2}+\frac{\partial^2 w}{\partial y^2}+\frac{\partial^2 w}{\partial z^2}\right) \qquad (4)$$

where ρ_f and μ_f are density and dynamic viscosity of the working fluid, respectively. Note, that p stands for pressure.

Energy Equation:

$$u\frac{\partial T_f}{\partial x}+v\frac{\partial T_f}{\partial y}+w\frac{\partial T_f}{\partial z}=\frac{k_f}{\rho_f C_{pf}}\frac{\partial p}{\partial z}+\left(\frac{\partial^2 T_f}{\partial x^2}+\frac{\partial^2 T_f}{\partial y^2}+\frac{\partial^2 T_f}{\partial z^2}\right) \quad (5)$$

Where $C_{pf,}$ k_f , and T_f are the constant specific heat capacity, thermal conductivity and temperature of the working fluid.

One of the important parameters in centrifugal compressor is stall margin, which can be calculated using the equation:

$$stall\ margin = 1-\frac{\dot{m}_{stall}}{\dot{m}_{design}} \qquad (6)$$

The total operating range for the performance map from stall to choke is calculated using the following equation:

$$Operating\ range = (\dot{m}_{choke}-\dot{m}_{stall})/\dot{m}_{choke} \qquad (7)$$

IOP Conf. Series: Materials Science and Engineering **899** (2020) 012010 doi:10.1088/1757-899X/899/1/012010

The Performance analysis of centrifugal compressor stage is assessed by performance parameters i.e. isentropic efficiency and pressure ratio, from inlet duct of impeller to diffuser outlet. Pressure ratio is evaluated using below equation:

$$Pr = \left[\int P_d * \frac{dm^{\cdot}}{m^{\cdot}}\right] / P_0 \quad (8)$$

Where P_d is the pressure at diffuser outlet, P_0 is pressure at inlet of compressor, $m^{\cdot}$ is the mass flow rate and Pr stands for total pressure ratio. Total to total isentropic efficiency can be found using Pr found in the above equation:

$$\eta_{tt} = T_i(Pr^{\frac{\gamma-1}{\gamma}} - 1)/(T_d - T_i) \quad (9)$$

Where η_{tt} stands for total to total isentropic efficiency, T_i stands for inlet temperature and T_d stands for temperature at diffuser outlet.

2.2. Geometric Parameterization in BladeGen

The primary step in pre-processing is to acquire an accurate geometric model of the compressor. For the test case compressor under examination, data has been taken from the German Aerospace Centre (DLR) and this data has been 3D modelled in BladeGen. This tool allows quick definition of radial blade flows and gives an important link between blade design and simulations.

For the test case study, the inlet duct, impeller and diffuser have been modelled as single domain. Only single domain has been considered for numerical simulations to reduce computational efforts required for the simulations as shown in Fig 1.

2.3. Computational Method for Meshing (Turbogrid)

After modelling centrifugal compressor in BladeGen, an H-grid mesh topology was created in ANSYS TurboGrid module for the entire fluid domain i.e. Inlet duct, Impeller and diffuser. TurboGrid is a meshing scheme for turbomachinery configurations, it selects automatic and refined mesh quality for most complex geometries creating hexahedral structured mesh. To study viscous sublayer, a very refined mesh has been created near the wall by providing adequate number of nodes near the boundary layer. For the better mesh quality, compressor computational domain is divided to H-grid topology with splitter blades arrangement. Design tip clearance for this case is variable tip-clearance of 0.5 mm(at blade leading edge),0.3 mm(at blade trailing edge). Mesh is generated using H-grid with 25 number of elements at the inlet and outlet each. A grid independency test executed using 8 distinct grid sizes for complete compressor. The grid with total number of mesh elements 560788 and total number of nodes 504345, has been found adequate as alteration in pressure ratio and isentropic efficiency was negligible in this range of elements as shown in Fig. 1 [16].

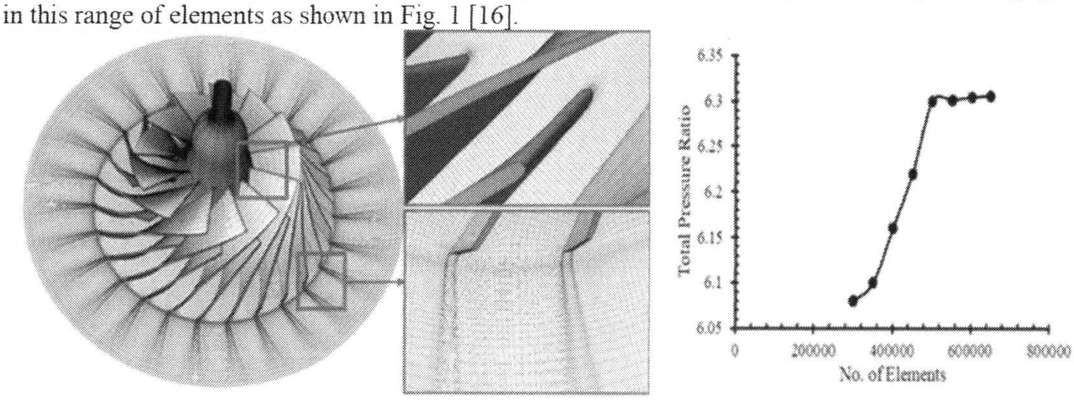

(a) Computational Mesh (TurboGrid) (b) Grid Independency Test

Figure 1: Computational Grid

2.4. *Turbulence Model*

For CFD simulations ANSYS CFX 15 was used to model the flow field for centrifugal compressor under steady state conditions. K-epsilon turbulence model is used for the numerical simulations. K-epsilon model gives better results for planar shear layer and recirculating flows as is the case of centrifugal compressor. According to the K-epsilon turbulence model criteria, the value of Y-plus is set 35.

2.5. *Numerical Setup and Boundary Conditions*

The compressor inlet conditions for all simulated cases.

$$T_i = 288.15\ k \tag{10}$$

$$P_i = 101.325\ kpa \tag{11}$$

$$Convergence\ Criteria = 1 * 10^{-5} \tag{13}$$

The medium turbulence intensity is kept 5% according to standard ambient conditions used by DLR (German Aerospace Center) for turbocharger configuration. The convergence criterion is set to the above mentioned value for all residuals while for monitoring convergence, mass flow rate and isentropic efficiency has been recorded for choke margin. As, at choke point minor variation in mass flow rate cause substantial variation in performance, so to check choke limit, static pressure is defined at outlet while mass flow rate is monitored, and choke point is found by slightly reducing static pressure at outlet [12]. The stage or mixing plane interfaces have been defined between diffuser, impeller and inlet. The rotational periodicity has been defined at boundaries of single passage assuming flow symmetry in all passages [17].

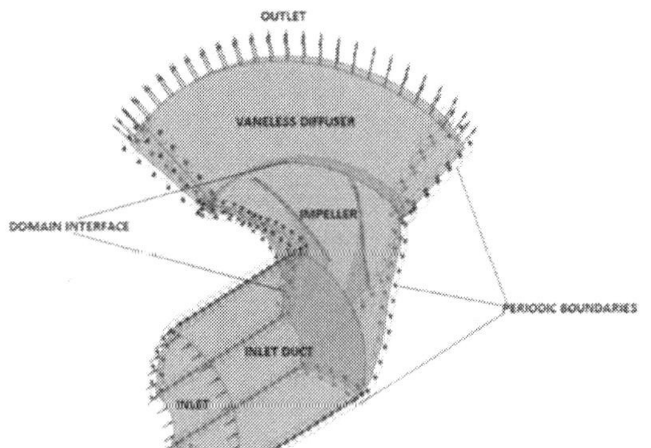

Figure 2: Computational Domain

2.6. *Creating the Design Space*

To study the effect of varying hub line, different design spaces have been created in the angle range of 40°-50° and the effect of the hub line variation has been analysed.

The impeller hub line has been modified using Bezier curve and analyzed its effect on the performance parameters of test case compressor as shown in Fig. 4. The new modified design has blade hub line adjusted at different angles from 40 to 50 degree, but the most optimized design has the angle adjusted at 45 degree as shown in Fig. 3. Bezier curves are used for describing the geometry of the curves.

$$p(t) = \sum_{i=0}^{n} ci\, B_i^n\,(t) \qquad t\epsilon[0,1] \tag{14}$$

Where ci represent control points and B_i^n represent ith Bernstein polynomial degree of n. To control the curve systematically, Bezier curves are used for this purpose. Which allow control of the curve and the curve obtained by the help of continuous derivatives is smooth line. Control polygon is the name given to control lines between successive control points. Bezier curve passes through the control points and specifically through first and last point. So, in this research project, hub line has been adjusted using Bezier curve technique.

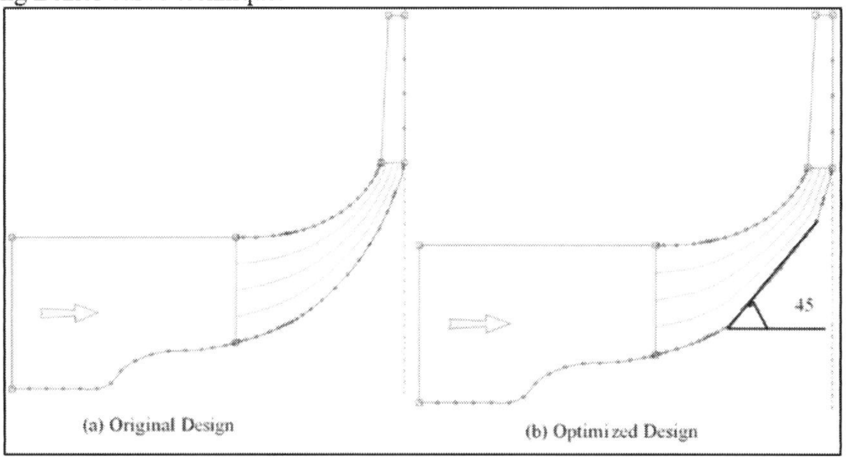

Figure 3: Design Space

3. Results and Discussions

3.1. Overall Performance

The overall performance of the compressor is analysed by making following agreements with the experimental test case:

- Design mass flow rate of 2.55kg/s
- Design speed of 50,000 rpm
- Turbulence model is k-ε model

Numerical simulations are carried out using ANSYS CFX solver at design conditions, total pressure ratio and total to total isentropic efficiency has been extracted from stall to choke to compare with the experimental data. The validation of the experimental data is carried out to show the suitability of this solver with the real-world problems. A very realistic results were shown by the simulated data as numerical simulations over predicted the experimental data by 9% only. Then numerical simulations are performed for the optimization of new compressor design as shown in Fig. 4. The optimization technique was performed by the hub line from 40° to 50° angle using Bezier Curve technique by keeping the rest of compressor same as experimental design. The analysis showed that hub line at angle of 45° angle shows the highest operating range and stall margin as shown in Fig. 3 and Fig. 4.

The operating range and stall margin for the experimental and numerically optimized design is calculated using equation 6 and 7. The operating range for experimental design is 16.9% while operating range for the CFD optimized design is found 18.18% from stall to choke. Hence, the total increase in the operating range for the optimized design is 7.5%. The stall margin for the experimental test margin is 5.7% while 8.24% for the optimized design. Hence, the percent increase in stall margin for the optimized design is 44% as shown in Fig. 4. All these compressors with high operating range and stall margin are used for the turbochargers applications to overcome the flow instabilities at low mass flow rate. The flow instabilities like flow recirculation and flow reversal due to adverse pressure gradient at low mass flow rate is called stall. The analysis only showed difference in the stall point, so the rest of analysis will be carried out at stall point to compare the experimental design and the CFD Optimized design.

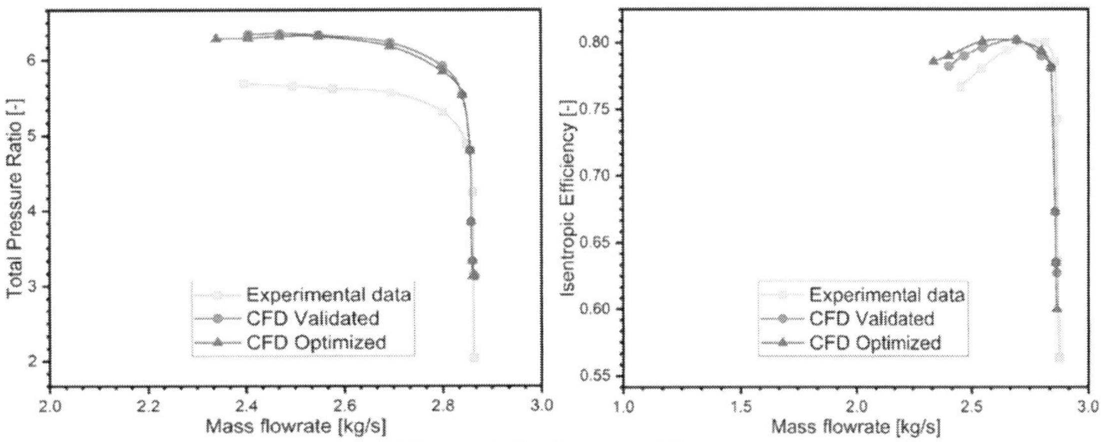

Figure 4: Performance Map

3.2. Diffuser Flow Separation

Factors that leads to centrifugal compressor instabilities are the flow separation and flow recirculation due to adverse pressure gradient, termed as Stall. These flow instabilities occur at low mass flowrate which leads to noise, vibration and sometimes fatal failure. In most of the cases flow separation occur at the trailing edge of the impeller blade and extends towards diffuser shroud, which has drastic effect on the compressor performance. These flow recirculation and flow separation effects is shown in Fig. 5, shown by the stagnant velocity contours and wakes at diffuser inlet. It is evident from the figure that these flow recirculation/wakes/jets are more obvious in the experimental design as compared to new optimized design. Which means that the new optimized design gives better performance to control flow instabilities and can be used in applications like turbochargers compressors.

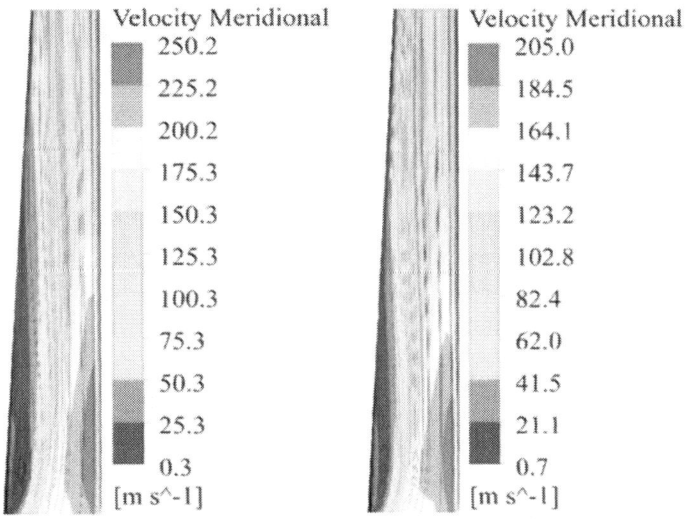

(a) Diffuser of Experimental Design (b) Diffuser of Optimized Design

Figure 5: Diffuser Flow Separation

3.3. Entropy Generation

In Fig. 6 and Fig. 7, entropy generation for stall point from inlet to outlet of the impeller along the meridional plane has been drawn which shows that entropy generation for the optimized design is smaller as compared to experimental design. This means that losses in the optimized design has been significantly reduced, which occurs because of secondary flow due to tip leakages, flow separation and flow recirculation. Then, static entropy contours are drawn along meridional length normalized and analyzed the maximum entropy regions, it is evident from the figure that maximum losses region is the blade trailing edge and it is significantly reduced in case of optimized design.

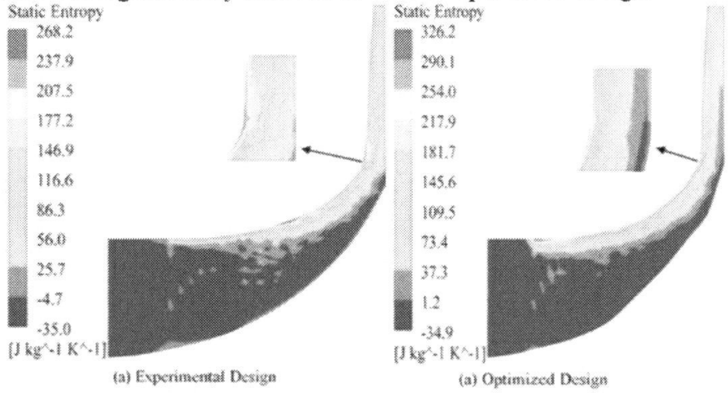

Figure 6: Entropy Generation from Inlet to Outlet of Impeller

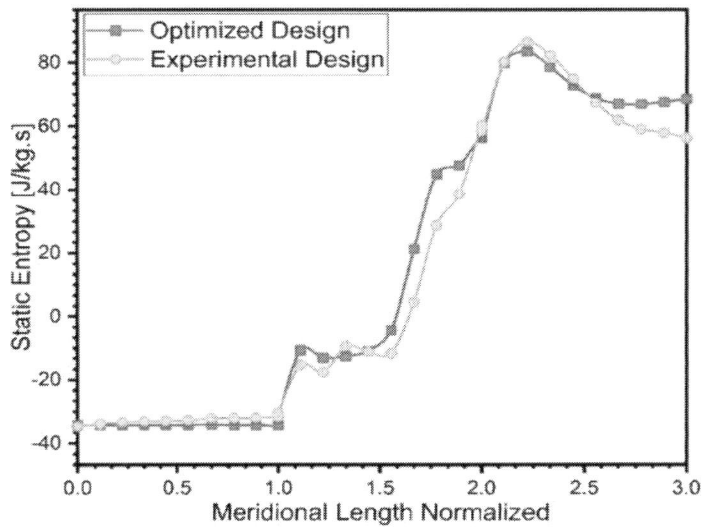

Figure 7: Entropy Generation from Inlet to Outlet of Impeller

3.4. Blade Loading

Blade loading represents the maximum load that is applied on the blade, which means that increasing blade loading causes more fatigue failure in case of centrifugal compressor. It is usually represented on the static pressure rise on each side of impeller blades (suction and pressure side). As it is evident from the Fig. 8, that blade loading for experimental design is higher as compared to optimized design for the same pressure ratio at stall point, which means that optimized design will have more fatigue life for the same application and same pressure ratio.

3.5. Meridional Velocity

Meridional velocity contours can be efficiently used to study the flow separation and shock waves generation in centrifugal compressor. It is clear from Fig. 9, that flow separation occurs at the diffuser shroud of the compressor and maximum velocity flow appears at impeller hub side. Meridional velocity is maximum at the hub line of the centrifugal compressor, which is reduced in the optimized design and as a result it reduces flow separation on the diffuser shroud. This means that tolerance capability for stall to control is maximum in optimized design. The figures indicate that at same pressure ratio and same efficiency, the optimized design will have small shock losses, small flow separation and low flow instabilities. So, the optimized design performance is much better than the experimental design while keeping meridional velocity, Mach number, total pressure ratio, isentropic efficiency and flow instabilities in view.

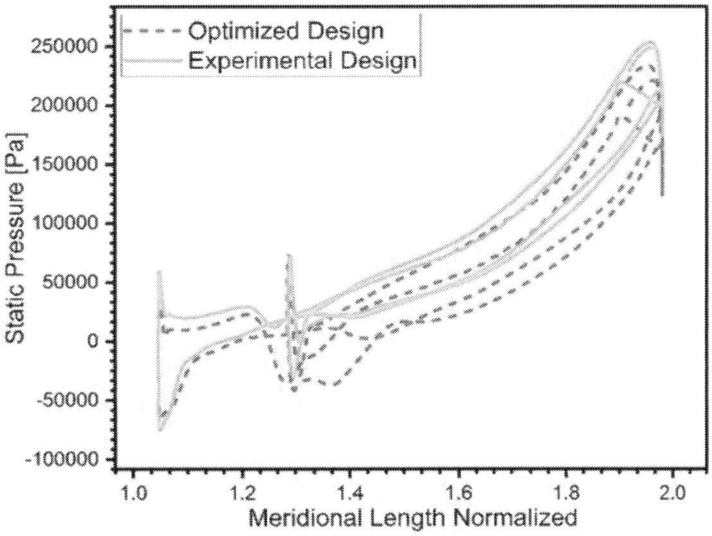

Figure 8: Blade Loading Diagram

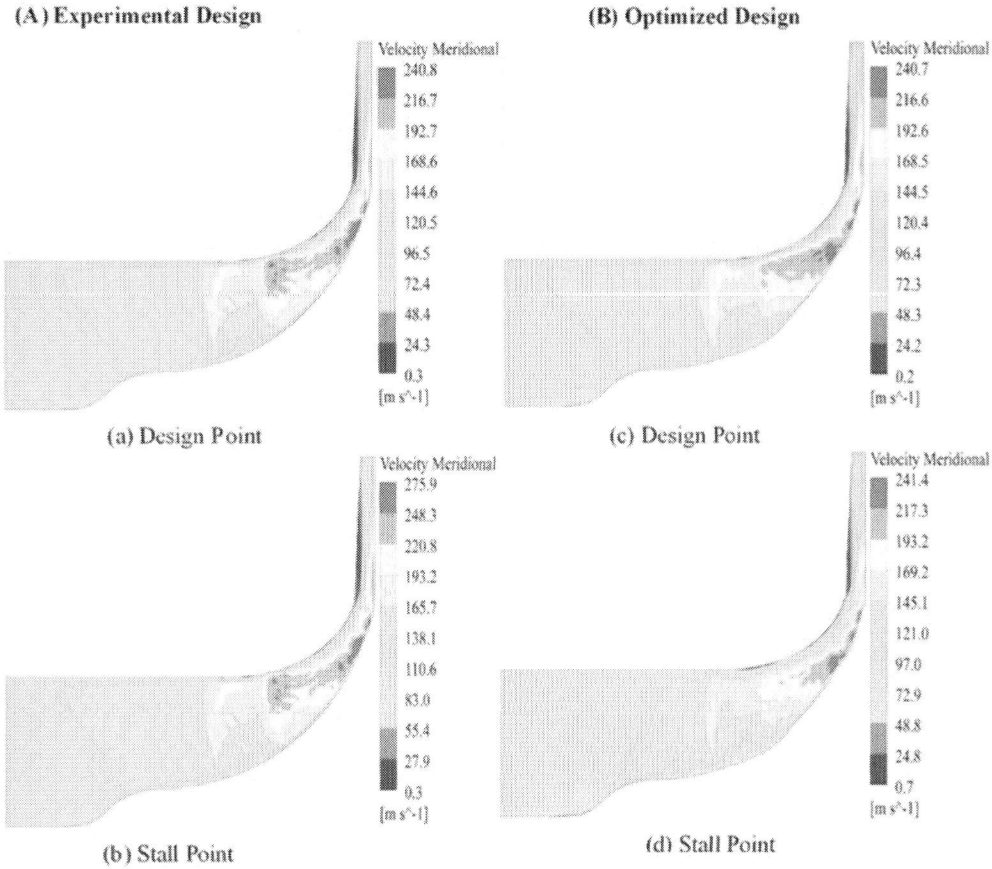

Figure 9: Meridional Velocity from Inlet to Outlet of Impeller

4. Conclusions

For the hub line optimization, 10 different designs have been analyzed by straitening the hub line from angle 40° to 50°. The 45° hub line is found the most optimized design with high operating range, stall margin, pressure ratio and isentropic efficiency. Based on detailed steady state computational fluid dynamics analysis, following conclusions can be drawn:

- As the hub line is straightened, made the impeller design much simpler. The simplified design results in low cost and low processing time
- Ansys CFX showed very realistic results by overpredicting the experimental data by 9% only
- By applying passive flow control method for flow instabilities, pressure ratio and efficiency decreases but in case of optimized design very slight reduction in pressure ratio and efficiency is found
- The operating range is increased by 7.5% while the stall margin is improved by 44% for the same operating conditions

Table II: Summary of Conclusions

Parameters	Experimental Design	Optimized Design
Operating Range (%)	16.9	18.18
Stall Margin (%)	5.7	8.24
Pressure Ratio	6.34	6.30
Isentropic Efficiency (%)	78	79

References

[1] A. R. Avio, "Aerodesign and Performance Analysis of a Radial Transonic Impeller for a 9 : 1 Pressure Ratio Compressor," vol. 115, no. July 1993, pp. 573–581, 2017.
[2] H. Xie, M. Song, X. Liu, B. Yang, and C. Gu, "Research on the Simplified Design of a Centrifugal Compressor Impeller Based on Meridional Plane Modification," *Appl. Sci.*, vol. 8, no. 8, p. 1339, 2018.
[3] C. Compressors, "Lesson 21," *Comp. A J. Comp. Educ.*, pp. 1–26.
[4] T. Ullah *et al.*, "Blade Meridional Profile Optimization for Novel High-Pressure Ratio Centrifugal Compressor Design Using Numerical Simulations," 2020.
[5] T. Shibata, M. Yagi, H. Nishida, H. Kobayashi, and M. Tanaka, "Performance improvement of a centrifugal compressor stage by increasing degree of reaction and optimizing blade loading of a 3D-impeller," *Proc. ASME Turbo Expo*, vol. 7, no. PART B, pp. 1305–1315, 2009.
[6] X. Xue, T. Wang, T. Zhang, and B. Yang, "Mechanism of stall and surge in a centrifugal compressor with a variable vaned diffuser," *Chinese J. Aeronaut.*, vol. 31, no. 6, pp. 1222–1231, 2018.
[7] A. Khan, M. U. Siddiqi, and F. Ahmad, "Numerical Investigation of Transonic High-Pressure Ratio Centrifugal Compressor," vol. 16, no. 2, pp. 83–95, 2020.
[8] A. Stein, S. Niazi, and L. N. Sankar, "Computational Analysis of Stall and Separation Control in Centrifugal Compressors Introduction," vol. 16, no. 1, 2000.
[9] A. Khan and A. Javed, "Performance Variation of Transonic Centrifugal Compressor at Different Tip Clearances Using Numerical Simulations," pp. 2–7.
[10] A. Khan and T. Ullah, "Numerical Analysis of Flow through High Pressure Ratio Centrifugal Compressor Impeller and Effect of varying Diffuser Exit Width on Performance," vol. 6, no. 11, pp. 399–405, 2019.
[11] C. Xu and R. S. Amano, "Meridional considerations of the centrifugal compressor development," *Int. J. Rotating Mach.*, vol. 2012, 2012.

[12] A. Javed and E. Kamphues, "Evaluation of the Influence of Volute Roughness on Turbocharger Compressor Performance From a Manufacturing Perspective," p. V02DT42A040, 2014.
[13] T. Raitor, O. Reutter, M. Aulich, and E. Nicke, "Aerodynamic design studies of a transonic centrifugal compressor impeller based on automated 3D-CFD optimization," *10th Eur. Conf. Turbomach. Fluid Dyn. Thermodyn. ETC 2013*, pp. 809–819, 2014.
[14] G. Eisenlohr, P. Dalbert, H. Krain, H. Pröll, F.-A. Richter, and K.-H. Rohne, "Analysis of the Transonic Flow at the Inlet of a High Pressure Ratio Centrifugal Impeller," p. V001T01A007, 2014.
[15] M. M. U. Rehman, T. A. Cheema, F. Ahmad, M. Khan, and A. Abbas, "Thermodynamic Assessment of Microchannel Heat Sinks with Novel Sidewall Ribs," pp. 1–12, 1981.
[16] P. Kapoor, "Influence of Surface Roughness on Flow Instabilities in a Centrifugal Compressor," pp. 1–24.
[17] F. Ghigliazza, A. Traverso, M. L. Ferrari, and J. Wingate, "Gt2008-50562," *Asme Gt*, pp. 1–9, 2008.

Home Security and Automation Based on Internet of Things: A Comprehensive Review

Saman Fatima[1, a], Naila Aiman Aslam[1, b], Iqra Tariq[1, c], Nouman Ali[1, d]

[1] Department of Software Engineering, Mirpur University of Science and Technology (MUST), Mirpur- 10250 (AJK), Pakistan

Email address: a) engr.samanfatima@gmail.com

Abstract. The current trends of research are focused on the use of the Internet of Things (IoT) for various real-time applications. IoT is a type of "universal global neural network" and very common because of its benefits in smart application trends over traditional communication trends. Information is now accessible much easily because of the internet and inexpensive. So security and automation become a need of every department and organization. The main objective of this study is to overcome the different face detection approaches in the IoT domain and their application in the automation of smart home devices. However, this review concludes the constraints and gaps of existing published work related to the current topic and point towards feasible future research directions. Furthermore, we have discussed possible challenges of large data processing, as well as communication and automation protocols by applying different sensors, other hardware, and their interoperability. Similarities and differences of various key factors are presented in tables in terms of the proposed model, technology, application, software, user interface, algorithm, hardware, and purpose, etc. In this study, we also discussed the pros and cons of various facial detection approaches and gaps of IoT in home automation. We suppose that this review will be a solid base for future researchers in the field of automation and facial detection in IoT.

Keywords—Internet of Things, Smart Homes, Home Automation, Security

1 INTRODUCTION

Nowadays, technology is persuading our lifestyle from the way we react to the way we behave and make it more attractive like the computer, smartphones, smart television, and cars. The usage of Internet of Things (IoT) is dramatically growing according to the business situation [1].

1.1 Internet of Things

IoT is known as a network of interconnected devices, embedded with software, network connectivity and sensors and electronic devices to gather and share information to make those devices sensitive [2]. From the air conditioner that we can control from our smartphone to the smart car providing to the shortest route or smartwatch tracking our daily activities, IoT is creating a giant network with connected devices. These smart devices are connected to each other and share data about how they are used and the environment in which they are operated [3]. IoT provides a collective databank to all devices to store their data and same language to all kind of devices for mutual communication. Data is sent from different sensors to the IoT platform. It accesses the received data of various sources. Then it analyse the data and extract the information according to the requirement. In the end, the resulted information is shared between the connected devices to this IOT platform for better user experience in the future. The commonly used framework of smart homes using IOT is shown in fig.1 [4]. It contains smart home devices that are connected with network gateway. The user can easily control all home appliances with the smart phone or tablet. Smart home automation saves the times and effort of the user\owner.

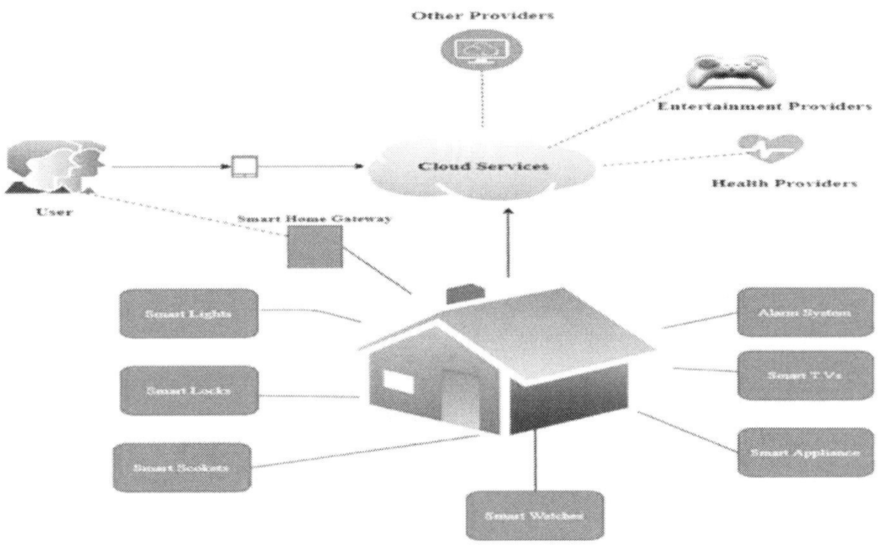

Figure 1. Commonly used framework for home automation [4]

The trend of establishing automatic connections between different devices is increasing dramatically day by day [5]. As shown in Fig.2 almost 15 billion devices were used to be connected in 2015 and just after 5 years, this number has been doubled in 2020 where around 31 billion devices are being connected. The incremental ratio of this trend is going to be above 1.5%in the upcoming 5 years until 2025.

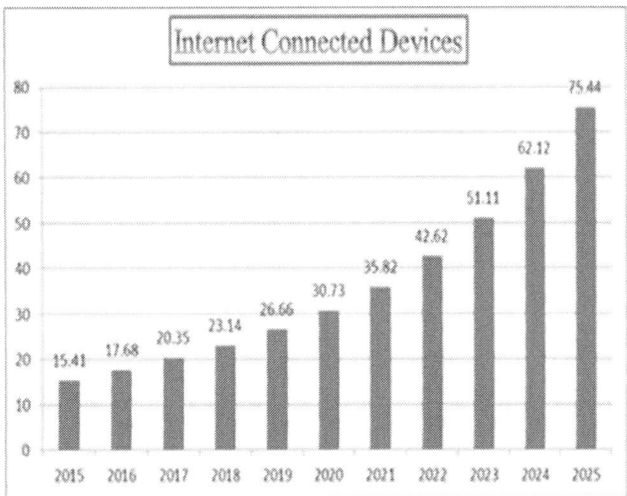

Figure 2. Incremental ratio of IoT devices in coming years

In the last few decades, security performs a very important role in our life. The safety of a house and households is the most important for everyone. In the past, people were used to have one pet at home for their security but now a day's technology has changed the situation [6]. Home automation has provided the security assurance even when we are away from home by giving us the access to control our devices at home via smart phone or tablet. It's highly strange for it to be a universal technology without describing the constraints and challenges in development of IoT and facilitating the privacy and security to the user. The distribution of IoT needs demanding efforts to implement and provide solutions for its security and privacy risks [7].

Face recognition system is used in security and protection of our property, companies, homes, markets and offices in the highest way as possible [8]. There are lots of problems and lacking in conventional safety systems because sometimes they do not provide data in real-time. Moreover, in biometric technology, a person can be recognized by using face recognition, which can extract different facial features of person formally and then recognize a person by applying different detection algorithms [2].

Security in home, apartments, companies and industries is become more essential now with the help of internet, is moved from traditional practices towards automation. Due to home automation, it not only gives comfort to human labour but also save the energy and time [9] [10]. Nowadays, home systems are automated and become smart and intellect with computing and information technology. It is necessary to make doors more secure because these are the main gateway to our home [11]. Currently using security systems are bare-metal locks and some other automated locking systems to secure doors. Performance and efficiency of smart locking systems can be evaluated in terms of cost. In recent years, different sensor devices are used to control home appliances and automated devices for home automation [12]. For handicapped and elder home members, home automation technology is much helpful. It helps to save energy of system sensors and also helpful for human labour. Therefore, big challenges for smart home systems are security and control management. Recognition based faces is most common biometric, which is used in control automation. Through smart mobile application based cloud services, e-mail or web based-services, a manager can access all the resources remotely, automatically and manually [6]. Face detection based on human discrimination in home automation is now an advance technology in which managers have an access to control all the resources remotely, naturally and manually through mobile application-based cloud Services [13] [14].In this is era of information, access to internet has become more reliable and easy. Internet has made our life secure, manageable and easier. Smart devices are now available, such as smart TVs, computers, tablets, smartphones and smart vehicles etc. Smart objects are able to perform different tasks, includes sharing information through web or app and generating alerts via internal or external sensors and turning on and off devices [15]. In other terms, by utilizing the IoT, smart homes can control and automate house lights, microwave, fridges, doors, multimedia, windows, and vehicle systems.

The contribution of review paper is to discuss the using of internet of things in home security and automation. The review paper gathered the recent publish work, summarize the recent work and analyse the different technologies use in the smart homes. The sections of the review paper are having been organized as follow: discuss the how IoT works in smart homes. Sections II describes literature review of the different technologies use in smart homes and draw the comparison table in sub headings and section III describes the conclusion and possible future works.

2 TECHNOLOGIES USE IN HOME SECURITY AND AUTOMATION

Wireless Home Security (WHS) and Home automation are the dual aspects of this research [16]. The main components of this research are TI CC3200 Launchpad, accessible Wi-Fi, Passive Infra-Red (PIR) motion detector sensor and relay for connecting the home appliance. If the security system detects any unexpected motion then the owner will receive a call on his mobile. The owner can turn ON lights and alarm. In the second module, when the guest teaches the owner's house then he will receive a video call. Now the owner can turn on the lights, fans, and A.C for his guest. The facial expressions are based drowsiness detection system using dynamic modelling. During drowsiness, techniques of computer

IOP Conf. Series: Materials Science and Engineering **899** (2020) 012011 doi:10.1088/1757-899X/899/1/012011

vision are used to detect and recognize the facial expressions, motion and changes in appearance of driver. Camera in front of the driver capture stream of images and sent to system as input. Input images then go through the four vital image processing stages. Overall process is as: 1) face detection and tracking 2) feature extraction 3) feature selection and 4) classification. Hidden Markov Models (HMMs) is used for drowsy and non-drowsy dataset. Author used ad boost cascaded classifier for comparing results. HMM-based dynamic modelling has batter results as compared to single frame-based technique. HMM-based results have 97% accuracy in classification and which is more than 90% accuracy, achieved by single frame-based drowsiness detection's house then he will receive a video call. Now the owner can turn on the lights, fans, and A.C for the guest [17].

2.1 Arduino-Based Smart Homes

Sawidin et al. [18] proposed home control system module is based on android. The main functions of this system are to control the lights, close and open the door, water faucet and garage of the home. The main components of this system are Ethernet shield, Arduino Uno, relay and android phones. When the user press the button open the door then door open and message send to the user. Asadullah et al. [19] introduced a smart home automation system using bluetooth technology. This system is the combination of hardware and software. Smartphones, the Arduino board and bluetooth modules are the main parts of hardware. Arduino integrated development environment (IDE) and bluetooth are parts of the software. Ultra-sonic sensor is also used to detect the water level inside the water tank as shown in Fig 3. When the water tank is filled then it automatically switches off the water pump and report is send to the smartphone. The soil moisture sensor is used to measure the water level inside the soil and also send a report to the smartphone. Arduino Uno is a microcontroller board which is based on A Tmega328P processor and bluetooth moduleHC-06 terminal used for wireless communication between the Arduino board and the smartphone. ASCII code is transmitted from smartphone to Arduino board via bluetooth terminal to control home appliances. The user will control the maximum of 18 appliances and sensors. Kak et al. [12] proposed a system for security purpose to control and manage the homes. The author develops the web application which run on cloud server and used by administrator anywhere, anytime to increase the security. The main objective of proposed system is to develop control management and automatic secure smart home system by using face detection and biometric techniques. The paper criteria certifying that proposed system is less expensive, user-friendly, adaptable and reliable for home members. The home owner can easily observe and control home activities by using web application. Only administrator can use the system and also can add, delete the members from dataset the proposed system also use PaaS (Platform as a Service). IaaS (Infrastructure as a Service) and SaaS (Software as a Service). Three types of member authentication process face detection, consecutive section, and feature extraction are used in this research paper. System components which are required for proposed system are Arduino microcontroller board, front door and back door camera, cloud devices, web based application.

GS,G et al. [20] proposed an Arduino-based home automation system. Arduino is connected to Wi-Fi and system is controlled by android application or social media network. The proposed system deals with the safety of home, and performs home technology with cost efficient. Different sensors are attached on Arduino which receive input signals from sensor. Analog temperature sensor chip as also used to tell about temperature conditions. This system is built for home automation. At the house door some motion sensors are fixed which detect any type of movement near the door. When sensor detect the darkness light 1 automatically on. The cooler or fan on when the room temperature exceed. The gas sensor MQ-6 is placed in the kitchen to detect any type of gas leakage in the kitchen. If gas leakage is detected the alarm which fix in the hall is raised. Relay is also used in this proposed system to switch electrical appliance light, fan, cooler etc. Intel Galileo is also placed in store room or garage which connected with Wi-Fi card with antennas for internet connectivity. System the intrusion detection featured also implemented which work with the help of motion sensors and control from web applications or android Applications. System components which are used in this system are Arduino, analog temperature sensor, DHT11 digital temperature humidity sensor, thermistor for the measurement of surrounding air, data pins, PIR sensor to sense the motion, MQ-6 gas sensor, relay, alarm, intel Galileo, motor drivers, power pins, and USB connectors.

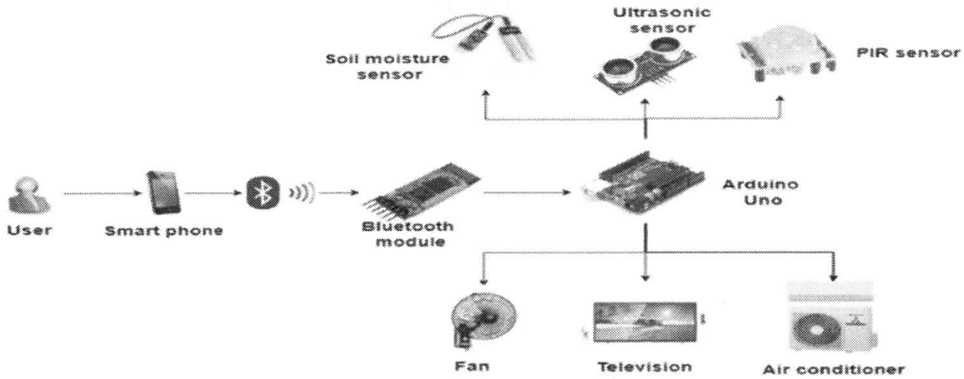

Figure 3. Arduino-based network [19, 20]

Kesavan et al. [21] proposed a system which used to automate home appliance. Home appliances are automatically ON and OFF. The proposed system also detects gas leakage and sends notification to home owner using the smart phone. This system cannot sense human existence and not provide lightening functionality which sense person is existing there or not. A detailed summary of above mentioned different Arduino-based home automation system shown in Table 1 after perform a review.

Table 1. SUMMARY OF ARDUINO-BASED SYSTEM

Author	Proposed Model	Hardware	Software	Platform	Purpose\ Application
AsadUllah et al. [19]	Bluetooth Arduino	Arduino module , bluetooth module, smart phone, soil moisture sensor ,ultrasonic range sensor	Bluetooth terminal smartphone application and Arduino Integrated development Application(IDE)	Smart phone application	Control home appliance , check water level inside the tank ,water level in the soil
Gunaseka ran et al. [20]	Wi-Fi based using Arduino	USB connector, power port, power pins, voltage regulator	Motion detection algorithm on Arduino	Web based application	Temperature, motion detection, fire detection, door, light level video monitoring, controlling appliances.
Chan, Kok et al. [22]	GSM based using Arduino	GSM module, relay 5 volt, cell phone, Arduino Uno	Python algorithm	Python supported mobiles	Control appliances
Zaw Lin et al. [23]	Bluetooth base using Arduino[24]	Arduino module, smart phone, soil moistening sensor, ultrasonic sensor	Python algorithm on Arduino module	Python supported mobiles	Secured and low cost
Kesavan et al.[21]	IOT framework for monitoring home automation	Arduino Uno R3, Ethernet, server, sound, gas and temperature sensor	Arduino IDE , android application, data filter algorithm	Linux, windows and smart phone platform	Monitor home status using mobile phone easily, automatically alert notification.

2.2 Raspberry Pi Based Systems

Raspberry Pi is circuit board or micro controller board having multiple components attached on it [25]. General purpose Input\Output (GPIO) pins are attached on board which connects the entire home appliance like LED, AC, fan, and television as shown in fig.4. These appliances are controlled by raspberry pi module on basis of different algorithms specified according to user requirements.

Figure 4. Raspberry Pi Based system

Othman et al. [26] embedded system is designed for human detection using smart security system. Images are captured by camera and sent on a smartphone. A possible solution to utilize Computer Vision techniques in the IoT. System is implemented using Raspberry Pi camera and a processor, known as Raspberry Pi 3. An image is captured by system when PIR detects any motion. For enhancement, computer vision techniques are applied on the captured images. Captured images then sent to smartphone through Wi-Fi or internet. Telegram application based on IoT, is used to check running activities and capture images, notifications and alerts. Raspberry Pi is a single.

According to Ibrahim Mohammad et al. [2] proposed system is a facial recognition system. System is implemented using python, open CV, Simple Mail Transfer Protocol (SMTP), e-mail server and raspberry pi processor. Raspberry pi processor is lighter and smaller in size. It can control power and efficiently work, by lower power consumption. System generates a security alert message according to authorized person's utilities. Pi camera takes images from dataset directory. After enhancement, train these images by assuming them as vectors and save them in trainer.yml file. However, capture a fresh face in camera for comparison. For trained images, recognizer will accept image by returning data, which includes id and profile of captured person. System will send an e-mail if, captured face is unrecognized.

Pawar et al. [27] stated that proposed system is for security purpose and based on IoT and facial recognition. In this system smartphones and web applications are used for monitoring the persons which enter in home and also decide who will enter in the home. In this system web camera is connected to the Raspberry Pi which contain PIR (Passive Infrared) sensor and ultrasonic sensor. When someone try to enter in home web camera capture image of person than real face recognition done by using LBP (Local Binary Pattern). If detected image match with home member images which are already in database than door will unlock, if image not match than door bell will ring. If unauthorized person try to break the door then red light turn ON alarm will raised and SMS and e-mail with unauthorized person image will be send to the home owner. The house owner can track all the activities which happening in house by using android and web application which are connected to Raspberry Pi processor using internet. The owner can also add new person's faces (e.g. guests) into the database by using android and web application. The proposed system consists of following components; a) Raspberry Pi which is main processor all other components are connecting on processor. b) PIR sensor used to sense the motion. c) Ultrasonic sensor used sound wave to measure the motion from distance. d) Stepper motor for door locking controlled by Raspberry Pi. e) Face recognition algorithm apply on Raspberry Pi. Android and web application's to access the system by using internet.

Sayem et al. [2] proposed a system for security purpose which is based on facial recognition security system using IoT. The main objective of this system is to develop the smart security camera with the help of Raspberry Pi to recognize authorized and unauthorized faces. If unauthorized face is detected than send email alert with image caption to the system owner. This system can implement face recognition from poor quality image and perform well over both authorized and unauthorized datasets. The technique which used for proposed is system is face recognition leverage technique which written in python language. The requirements which are used to build the system are Raspberry Pi model B+, Raspberry Pi camera module V2, python programming language, software library Open CV (Open source Computer Vision Library) which provide infrastructure for computer vision, and SMTP (Simple Mail Transfer Protocol) for sending and receiving email.

Jain et al. [11] system is combination of doorbell interface and Raspberry Pi. When someone presses the doorbell the camera is activate and capture the face after this it checking with already registered faces in database. If the person is authorized than door will open. If the face does not match with database it send an email to owner with that burglar image and OTP (One Time Password). When unauthorized person type OTP provided by the owner knowledge than it will allow to enter in home. Different types of hardware are used in these systems which are Raspberry pi, SD card, PIR (Passive Infrared) sensor, camera, DC motor, push button. Software components which are used in this system are Raspbian Jessie operating system, python, open CV, Apache (server), and PHP for database connectivity.

Othman N. A. et al. [28] introduced the security of a house by using the IoT concept. It is extremely helpful and vital if we want to protect the area. The main circuit board is Raspberry pi attached with the pi camera, PIR and smart mobile phone. In this system, the camera has been used to take the pictures when a motion is detected via the PIR sensor. The system will not go for face detection if there is no motion detected. But, if any movement is detected, then the detected movement of the current frame will be processed by the Viola Jones algorithm of face detection and the resulted image will be sent and displayed on the smartphone. Harr, Ada Boost and cascade classifier are used for the classification of random pictures. The flow chart of this proposed system is given in Fig.5.When motion detects pi camera capture the image and detects face. Finally it saves the image and sends to the owner.

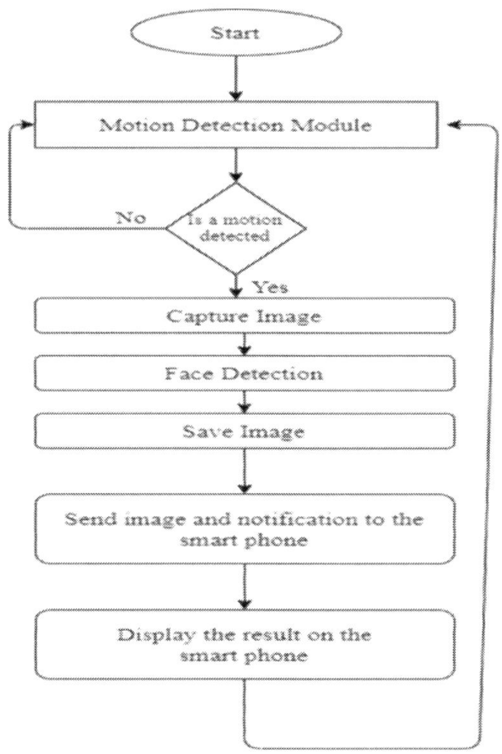

Figure 5. Flow chart motion detection and display result [28]

S.Tanwar et al. [6] proposed IoT based security alert system for the smart home. A PIR sensor and web came is used to detect and capture the image of any movement and deliver this image to Raspberry PI. In this research, the Pseudocode Approach is used. Signals of the PIR sensor are received from the GPIO pin and these signals are further connected with the Raspberry Pi. It takes the two signals from the PIR sensor and compares both of them with each other and if there is any differentiation between them then it registers it as motion and gives a command to the camera to start capturing the images. These images are stored on the temporary memory storage of Raspberry Pi and then it sends a defined mail to the user or the owner of the house with fixed subjects. A detailed summary of above mentioned different Raspberry Pi-based home automation system shown in Table 2 after perform a review.

Table 2. SUMMARY OF RASPBERRY PI-BASED SYSTEM

Author	System	Hardware	Software	User interface	Application	Benefits
Jain, A et al. [11]	IoT based smart doorbell using Raspberry Pi	Raspberry Pi, SD card, PIR sensor, camera, doorbell, DC motor, push button	Raspbian OS, python, open CV, apache server, PHP for database	Web based application or android	Face detection, capture image, send mail to owner	Low cost, Home security,
Pawar et al. [27]	IoT based home security and face recognition using Raspberry Pi	Raspberry Pi, PIR sensor, ultrasonic sensor, stepper motor, camera, alarm,	Face recognition algorithm	Web and android applications	Facial recognition, motion detection, image capture-e-mail send,	Convenience, safety and security
Sayem et al. [2]	Integration of face recognition security system with IoT using Raspberry Pi.	Raspberry Pi model B+, camera module V2, SMTP alarm	Face recognition algorithm using python	Mobile applications	Face detection, send e-mail with image,	Home security, Low cost, efficiency
Othman et al. [28]	Smart door security application using IoT	Raspberry Pi ,PIR sensor , Pi camera ,SD card	Viola jones algorithm	Android application	Face detection, image capture.	Home security
Tanwar et al. [6]	Smart Home automation using IoT	Raspberry Pi ,PIR sensor , Pi camera ,SD card	Pseudo code	Web and smart phone	Video capture, face detection, send email	Safety and security, low cost.

2.3 Protocols-Based system

Message Queuing Telemetry Transferring protocol (MQTT) is a light weight IOT protocols [29] [30].MQTT is a based on publish\subscriber [31] .Publisher publish the topic and broker receive the name of topic. When any client or subscriber need this topic it request to broker and broker send this topic to client. In MQTT, the broker is responsible of gathering data from various clients or subscriber. In fig.5 MQTT establish connection, maintain and terminate the connections. In which PUBLISH, PUBACK, PUBREC, CONNECT, CONNACK, PINGREQ, DISCONNECT, PUBREL, and PUBCOMP are the MQTT packets.

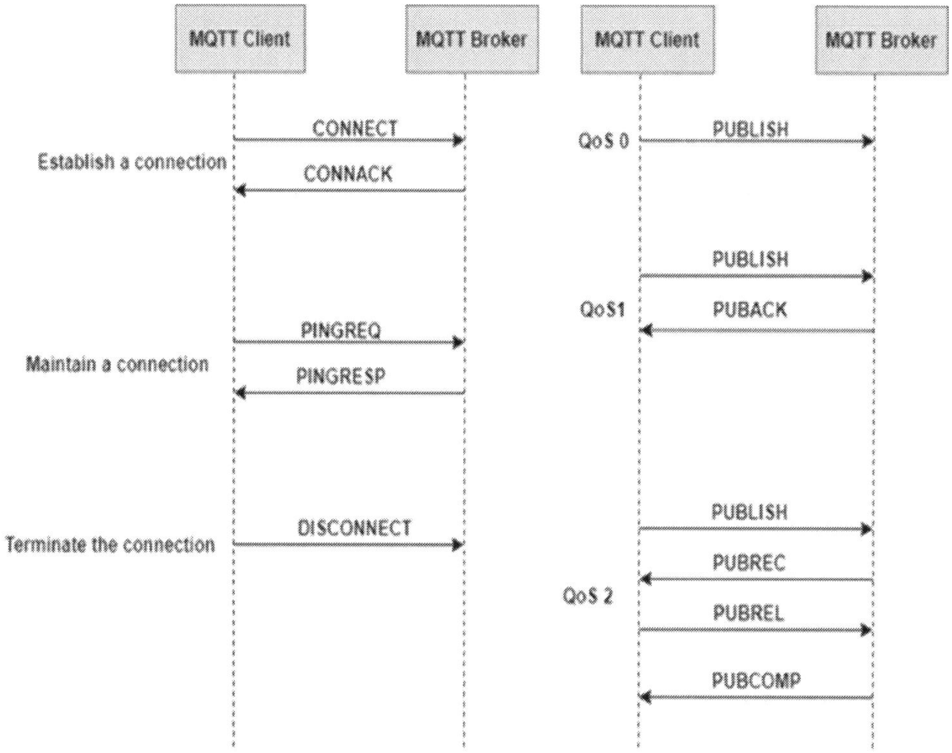

Figure 6. MQTT connections

Prabaharan et al. [32] proposed home security and automation system by using MQ Telemetry Transport (MQTT) protocol. The system can control gas leakage, turning Off\On lights or fans and temperature of home. In the system, the three main modules is used in the system, first module is based on temperature sensor, light dependent resistor and grove gas sensor. The second and third module is node microcontroller unit (MCU) and end user interface module called Ad fruit open source cloud platform.

Lohokare et al. [33] proposed ecosystem of home automation system by using MQTT, Kafka and Spark Streaming. MQTT is IoT enabled protocol has three main components subscriber, publisher and broker. The publisher devices generate data and publish this data via topic. The broker server receives data in the form of topics and sent this data to the relative subscriber. When new device registered in the system then device get new id .When the temperature of device is greater than fifty then it send email to authorized user. Spark streaming analyse the data of the devices in every two minutes. A detailed summary of above mentioned different protocols-based home automation system shown in Table 3 after perform a review.

Table 3. SUMMARY OF PROTOCOLS-BASED SYSTEM

Author	Proposed Model	Hardware	Software	User Interface	Purpose\ Application
Prabaharan et al. [32]	MQTT protocol	Node MCUs ,buzzer alarm ,temperature and humidity sensor ,grove gas sensor, light dependent resistor	Ada fruit open source cloud platform	Mobile phones, tabs , Laptop	Control leakage of gas ,Light or fan On\Off ,check temperature and humidity
LohoKare et al. [33]	IoT protocols	Arduino enabled ESP8266, light intensity sensor , temperature sensor	MQTT broker, apache kafka ,apache spark streaming	Android application	Home automation
Kodali et al. [29]	MQTT protocol	ESP8266 processor	MQTT lens, MQTT broker	Smart phone, laptops	Home appliance
Upadhyay et al. [34]	MQTT protocols	Raspberry pi B+, apache server. humidity sensor , temperature sensor	MQTT-mosquito	Webpage	Home appliance

3 CONCLUSION

This review-based research addresses the vision about shift of traditional technology into automated technology in home automation. In this review, we have selected most relevant papers and articles through different search quires that address this topic comprehensively. Selected papers then divided into three different categories based on technologies used in home automation. First category is Arduino-based smart homes, in which we have discussed Arduino-based papers and comparison of those papers in tabular form including proposed model, hardware, software, user interface and purpose. Second category is about raspberry based home automation, in which we have discussed papers based on raspberry pi processor. Further we have also made a comparison of reviewed papers in tabular form including hardware, system, software, user interface, application and benefits. In third category, we have discussed papers based on protocols used in home automation and comparison of those papers in tabular form including proposed model, hardware, software, user interface and application.

Where IoT provides us benefits, it also has some issues. It is a complex network where a minor failure can cause serious consequences. Compatibility issues are also there because different devices from different manufacturers are interconnected. The main limitation of IoT is technology has taken control of life and we are now fully dependent on it.

In future, work can be done on face detection system based on state-of-art, IoT security, block chain-based IoT, Mobility in IoT security and automation of routers and integrated artificial intelligence (AI).

4 References

[1] Miraz, M.H., et al. A review on Internet of Things (IoT), Internet of everything (IoE) and Internet of nano things (IoNT). in 2015 Internet Technologies and Applications (ITA). 2015. IEEE.
[2] Sayem, I.M. and M.S. Chowdhury. Integrating face recognition security system with the

internet of things. in 2018 International Conference on Machine Learning and Data Engineering (iCMLDE). 2018. IEEE.

[3] García, C.G., D. Meana-Llorián, and J.M.C. Lovelle, A review about Smart Objects, Sensors, and Actuators. International Journal of Interactive Multimedia & Artificial Intelligence, 2017. **4**(3).

[4] Ali, W., et al. IoT based smart home: Security challenges, security requirements and solutions. in 2017 23rd International Conference on Automation and Computing (ICAC). 2017. IEEE.

[5] Shahid, N. and S. Aneja. Internet of Things: Vision, application areas and research challenges. in 2017 International Conference on I-SMAC (IoT in Social, Mobile, Analytics and Cloud)(I-SMAC). 2017. IEEE.

[6] Tanwar, S., et al. An advanced internet of thing based security alert system for smart home. in 2017 International Conference on Computer, Information and Telecommunication Systems (CITS). 2017. IEEE.

[7] Farooq, M.U., et al., A review on internet of things (IoT). International Journal of Computer Applications, 2015. **113**(1): p. 1-7.

[8] Sánchez-Arias, G., C.G. García, and B.C.P. G-Bustelo, Midgar: Study of communications security among Smart Objects using a platform of heterogeneous devices for the Internet of Things. Future Generation Computer Systems, 2017. **74**: p. 444-466.

[9] Chowdhry, D., R. Paranjape, and P. Laforge. Smart home automation system for intrusion detection. in 2015 IEEE 14th Canadian Workshop on information Theory (CWIT). 2015. IEEE.

[10] Skeledzija, N., et al. Smart home automation system for energy efficient housing. in 2014 37th International Convention on Information and Communication Technology, Electronics and Microelectronics (MIPRO). 2014. IEEE.

[11] Jain, A., et al. IoT-based smart doorbell using Raspberry Pi. in International Conference on Advanced Computing Networking and Informatics. 2019. Springer.

[12] Kak, S.F. and F.M. Mustafa. Smart Home Management System Based on Face Recognition Index in Real-time. in 2019 International Conference on Advanced Science and Engineering (ICOASE). 2019. IEEE.

[13] Liu, Z., L. Lv, and Y. Wu. Development of face recognition system based on PCA and LBP for intelligent anti-theft doors. in 2016 2nd IEEE International Conference on Computer and Communications (ICCC). 2016. IEEE.

[14] Lwin, H.H., A.S. Khaing, and H.M. Tun, Automatic door access system using face recognition. international Journal of scientific & technology research, 2016. **4**(6): p. 210-221.

[15] Othman, N.A. and I. Aydin. A new IoT combined body detection of people by using computer vision for security application. in 2017 9th International Conference on Computational Intelligence and Communication Networks (CICN). 2017. IEEE.

[16] Kodali, R.K., et al. IoT based smart security and home automation system. in 2016 international conference on computing, communication and automation (ICCCA). 2016. IEEE.

[17] Tadesse, E., W. Sheng, and M. Liu. Driver drowsiness detection through HMM based dynamic modeling. in 2014 IEEE International conference on robotics and automation (ICRA). 2014. IEEE.

[18] Sawidin, S., D.S. Pongoh, and A.A.S. Ramschie. Design of Smart Home Control System Based on Android. in 2018 International Conference on Applied Science and Technology (iCAST). 2018. IEEE.

[19] Asadullah, M. and K. Ullah. Smart home automation system using Bluetooth technology. in 2017 International Conference on Innovations in Electrical Engineering and Computational Technologies (ICIEECT). 2017. IEEE.

[20] GS, G., et al., Traffic Vehicular Communication With Zigbee Protocol. Journal of Chemical and Pharmaceutical Sciences ISSN. **974**: p. 2115.

[21] Kesavan, G., P. Sanjeevi, and P. Viswanathan. A 24 hour IoT framework for monitoring and managing home automation. in 2016 international conference on inventive computation technologies (ICICT). 2016. IEEE.
[22] Teymourzadeh, R., et al. Smart gsm based home automation system. in 2013 IEEE Conference on Systems, Process & Control (ICSPC). 2013. IEEE.
[23] Oo, Z.L., T. Win Lai, and M. Than, Bluetooth Based Home Automation System Using Android and Arduino. 2018.
[24] Sriskanthan, N., F. Tan, and A. Karande, Bluetooth based home automation system. Microprocessors and microsystems, 2002. **26**(6): p. 281-289.
[25] Patchava, V., H.B. Kandala, and P.R. Babu. A smart home automation technique with raspberry pi using iot. in 2015 International conference on smart sensors and systems (IC-SSS). 2015. IEEE.
[26] Othman, N.A. and I. Aydin. A face recognition method in the Internet of Things for security applications in smart homes and cities. in 2018 6th International Istanbul Smart Grids and Cities Congress and Fair (ICSG). 2018. IEEE.
[27] Pawar, S., et al. Smart Home Security Using IoT and Face Recognition. in 2018 Fourth International Conference on Computing Communication Control and Automation (ICCUBEA). 2018. IEEE.
[28] Aydin, I. and N.A. Othman. A new IoT combined face detection of people by using computer vision for security application. in 2017 International Artificial Intelligence and Data Processing Symposium (IDAP). 2017. IEEE.
[29] Kodali, R.K. and S. Soratkal. MQTT based home automation system using ESP8266. in 2016 IEEE Region 10 Humanitarian Technology Conference (R10-HTC). 2016. IEEE.
[30] Singh, M., et al. Secure mqtt for internet of things (iot). in 2015 Fifth International Conference on Communication Systems and Network Technologies. 2015. IEEE.
[31] Hunkeler, U., H.L. Truong, and A. Stanford-Clark. MQTT-S—A publish/subscribe protocol for Wireless Sensor Networks. in 2008 3rd International Conference on Communication Systems Software and Middleware and Workshops (COMSWARE'08). 2008. IEEE.
[32] Prabaharan, J., et al. Wireless home automation and security system using MQTT protocol. in 2017 2nd IEEE International Conference on Recent Trends in Electronics, Information & Communication Technology (RTEICT). 2017. IEEE.
[33] Lohokare, J., et al. An IoT ecosystem for the implementation of scalable wireless home automation systems at smart city level. in TENCON 2017-2017 IEEE Region 10 Conference. 2017. IEEE.
[34] Upadhyay, Y., A. Borole, and D. Dileepan. MQTT based secured home automation system. in 2016 Symposium on Colossal Data Analysis and Networking (CDAN). 2016. IEEE.

ACKNOWLEDGMENT

The authors would like to thank the Department of Software Engineering at MUST University for facilitating the development of the paper and making available resources for final development of paper.

IOP Conf. Series: Materials Science and Engineering **899** (2020) 012012 doi:10.1088/1757-899X/899/1/012012

Investigation of Sensors & Actuators based on Hankel Singular Values

Dr. Sheharyar Malik[1]

[1] Instructor, Abu Dhabi Polytechnic, IAT, UAE.

sheharyar.malik@adpoly.ac.ae

Abstract. The work presented here proposes a method to rank the efficiency of the actuators and sensors for the attenuation of excited structural modes. Aeroelastic wing designed with multiple control surfaces and equipped with sensors at different locations is used as a test case. A unifying theme to this work is the application of Hankel Singular Values (HSV), an analytical approach applied to the experimental data conducted during the Ground Vibrational (GVT) and Wind Tunnel (WT) tests. Modal information gathered from experimental results is used to define actuator/sensor relationships, once the mode of interest is decided then the sensor and actuator location is determined to effectively cater the structural vibrations, it paves the way for the design of specific performance metrics. The research will add redundancy to the available set of instrumentation by paving the way for the dedicated use of control surface for the stable flight mechanics or flight dynamics instability. The results presented here are supported by Power Spectral Densities (PSD's) and band pass filters.

1. Introduction

Multi-surface control is used for static and dynamic aeroelastic tailoring of the wing where, these surfaces can be used to alter the camber of the wing by redistributing the aerodynamic load. One beneficial feature of the multi-surface control wing is that it allows the freedom of multitasking for various purposes like flutter suppression, load attenuation or flight mechanics, these possibilities make multi surface wing an attractive choice for aeroelastic wing. Multiple control surfaces offer many ways in which these surfaces can be instructed i.e. All the surfaces can be commanded at once (one surface) or different surfaces can be made to work independently of each other. Constrained optimization is helpful in maximizing the effectiveness of the of a control surface for the given mode [1].

The availability of the multi surface analysis provide redundancy and helps to rank the suitability of the control surface for any structural mode attenuation and to reduce the control system design from multi-input multi-output system to single-input single-output system. Advanced wings that employ multiple control surfaces are used to enhance their aeroelastic response under the projects of NASA Active Flexible Wing (A.F.W.) and Active Aeroelastic Wing [2, 3, 4]. Optimized placement of sensors with Identical Location of Acceleration and Forces (ILAF) technique enables to avoid spill over into other modes by focusing on modes of interest [5]. Work is presented for the optimal location of the sensors and actuators in the preliminary design phase, it is cost effective as well as energy efficient solution to the attenuation of the structural modes. The work relies on the novel usage of Hankel

Singular Values (HSV) which define the relation between inputs and outputs of the system in the time domain for lightly damped structures [6].

The work presented here also uses the HSV to rank the outputs and inputs for the associated excited structural modes and their corresponding damping. The aeroelastic wing has four control surfaces, two control surfaces are located at the leading edge and are named as Leading edge Outboard (LEO), Leading Edge Inboard (LEI). Similarly, two control surfaces are present at the trailing edge and are termed as Trailing Edge Outboard (TEO) and Trialing Edge Inboard (TEI). The accelerometers attached by following the ILAF approach provides the structural accelerations.

The following section II presents the analytical approach to the problem formulation, followed by the HSV analysis of experimental results in the Section III.

2. HSV Analytical Modelling

The analytical evaluation of the performances of each control surface is implemented by keeping in view the worst-case scenario, where one or more control surfaces is not functional and not getting the sufficient performance from the control surface. It can be simply inferred as quantifying the suppression of vibrations with respect to the energy/cost related to the actuation of the surface.

A unifying theme to this research is the usage of Hankel singular values (HSV) for quantitative attenuation analysis of actuator and sensor combination. This scheme is originally presented for optimized placement of actuator/sensor in the preliminary design phase. As the wing is already manufactured with predetermined location of instrumentation, HSV is used experimentally for integrated design perspective for this research. For lightly damped structure, its modal properties can be used to obtain HSV for modal coupling at preliminary design phase, given by the expression:

$$\sigma_{yui}^2 = \frac{[b_i^T b_i][c_i^T c_i]}{(4\zeta_i\omega_i)^2} \tag{1}$$

With 'b_i' as the input matrix and 'c_i' as output matrix of the i-th mode for the state space model. The numerator provides the coupling of actuator and sensor for the respective mode. Denominator defines the time constant of the state in terms of damping ratio 'ζ' and excited frequency 'ω' of the i-th excited mode as:

$$\tau_i = \frac{1}{(\zeta_i\omega_i)} \tag{2}$$

As for the presented case the data is preferred to be analyzed in the frequency domain. Also, the task on hand is to compare the efficient attenuation which includes the comparison of structural vibrations for active and inactive control system along with taking into consideration the optimized performance of the actuator. So, the task on hand is now to find the suitable combination for modal attenuation, the equation is transformed to find the effective coupling of actuators and sensors to damp out the first bending and first torsion modes under control law which minimizes the variance of actuator movement:

$$\sigma_{yui}^{\ 2} = \frac{(b_{i_{on}})(c_{i_{on}} / c_{i_{off}})(c_{i_{on}})}{(4\xi\omega_n)^2} \tag{3}$$

Where 'b_i' and 'c_i' represents the variance extracted from actuator inputs and structural outputs, subscripts 'on' and 'off' represents the readings taken when the control system is active or inactive. For the predetermined actuator sensor locations, the optimum coupling path of actuator signal and sensor output between each mode N_m is given as:

$$\sigma_{yu} = \mathrm{diag}(\sigma_{yu1}, \ldots, \sigma_{yun}, \ldots, \sigma_{yuNm}) \tag{4}$$

[7], suggested to improve the performance of control, by coupling each mode with performance-disturbance path as:

$$\sigma_{zw} = \mathrm{diag}(\sigma_{zw1}, \ldots, \sigma_{zw\,n}, \ldots, \sigma_{zw\,Nm}) \tag{5}$$

In the design metric form, it is given as:

$$J_{qp} \equiv \sum_{i=1}^{N_m} \frac{\sigma^2_{y_p u_q i}}{\bar{\sigma}^2_{yui}} \sigma^2_{zw\,i} \tag{6}$$

The essence of this task is to select the best couple of q-th sensor and p-th actuator for increase performance and efficiency. $\bar{\sigma}^2_{yui}$, defines the square of all HSV values for actuator/sensor coupling. A similar approach has been adopted to minimize the variance of the outputs: showed that performance 'z' can be controlled by introducing the weighting function in the method with systems performance. Design metric can be utilized to switch between the robustness and performance based cost functions.

3. HSV Experimental Analysis

An experimental setup was prepared to equip the wing with four PCB monoaxial accelerometers with bandwidth ranges from 0.5 Hz to 3 kHz. Two accelerometers are located near the wing tip on either side of the elastic axis and two accelerometers are located on the midspan of the wing. The following figure 1, shows the sensor/actuator schematic of the wing with symbolic representation of sensors.

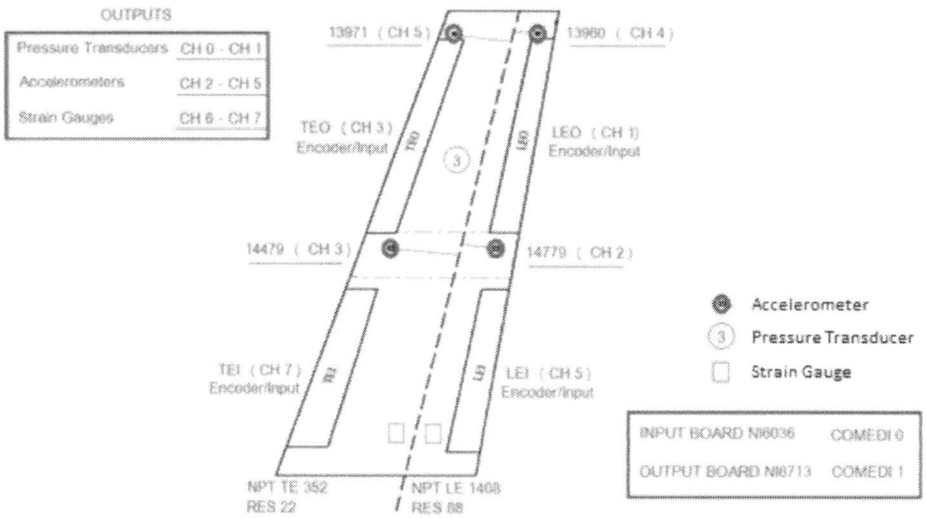

Figure 1. Instrumentation of the wing

The four control surfaces are equipped with the electric motors Portescap mod 17N78-210E for driving these control surfaces (LEI, LEO, TEI and TEO). are selected by fulfilling the criteria set by physical design constraints such as weight and maximum allowable size of the motors [18]. Each motor is also equipped with encoder. Thanks to planetary gears in shaft/line torque transmission, chief concern of torque and gear reduction are met. The signals from the sensors are filtered by KEMO antialiasing filters while the connectivity between the host and the target system is defined by the Real Time Application Interface (RTAI), the system provided the real time data processing as well as data was stored to be processed at the later stage.

IOP Conf. Series: Materials Science and Engineering **899** (2020) 012012 doi:10.1088/1757-899X/899/1/012012

The following figures shows the process of experimentally extracting variance and Hankel singular values for already assigned actuator/sensor locations. The PSD's from each sensor configuration are shown in the Figure 2 to extract the variance of sensor readings in the bending mode, the signals are filtered with 6th order equiripple filter so that the signal contains only one mode.

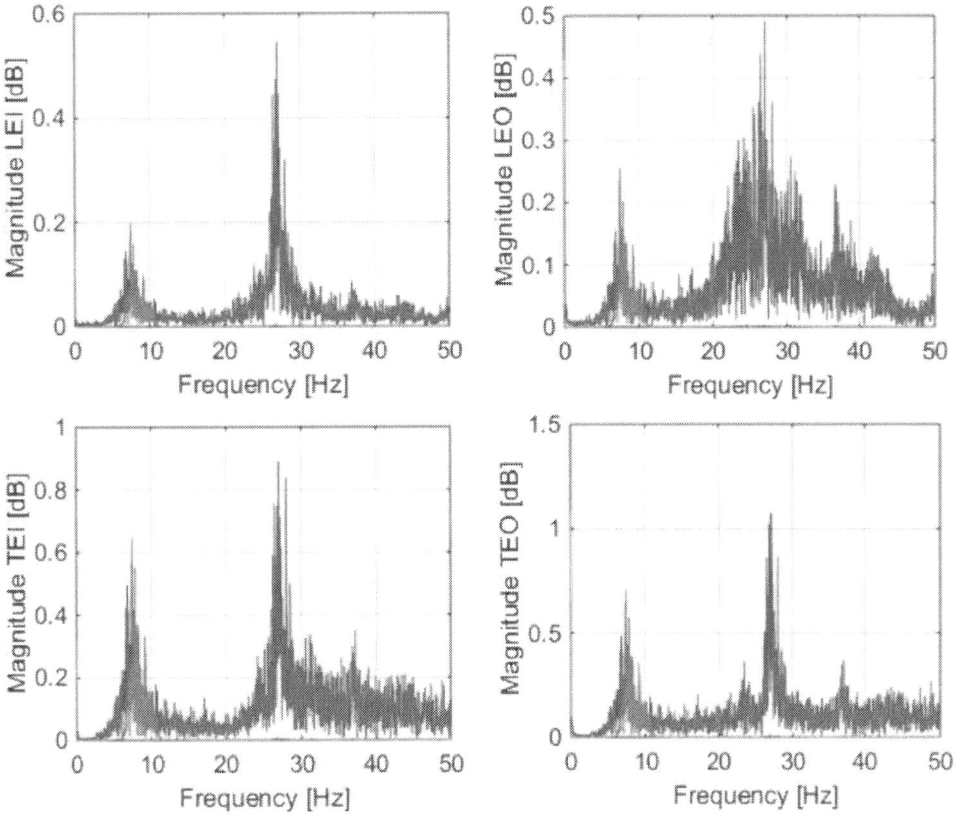

Figure 2. Extraction of Bending Mode – Control Off

Same procedure is followed to extract the variance for the bending signals when the control system is active (On) as shown in figure 3. Figure 4 shows the PSD of the attenuated and filtered modes under active control. The effect of active control system is evident by observing the reduction in structural vibrations.

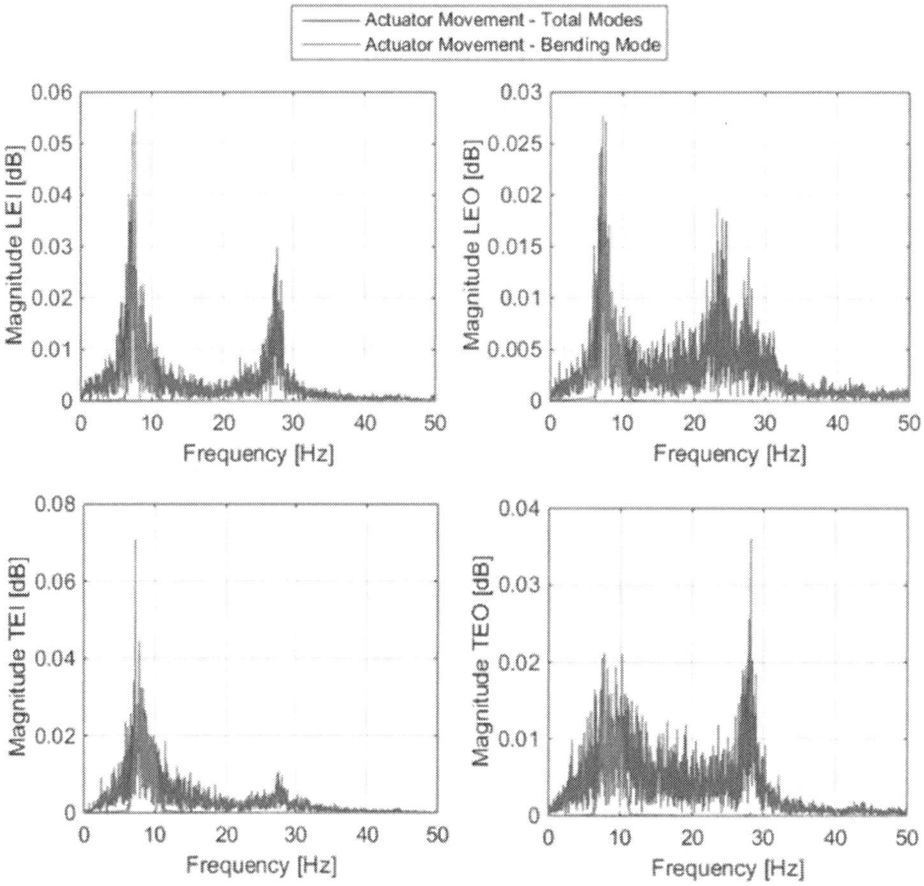

Figure 3. Extraction of Bending Mode – Control On

Figure 4, shows the PSD of actuator movement for active control during the attenuation various modes of bending mode. PSD of actuator movement in bending mode is realized also through equiripple filter. The extracted PSD is used to calculate the variance of the actuator movement in the bending mode. The Hankel singular Values can be calculated to show the effect of active control under optimized actuation movement for all the control surface working under ILAF concept. The values are analytically calculated by the help of the formula described in equation (6). It shows that most effective surface for bending mode is trailing edge outboard and most suitable surface for torsion mode is trailing edge inboard. Table 1 is mentioned below to signify the importance of each actuator and significance of actuator/sensor combination for cost function is, it can be readily seen that outboard trailing edge is more effective in bending mode participating frequencies while inboard trailing edge is more effective in torsional mode attenuation. Moreover, it is also noted the deficiency of leading edge strip i.e. LEO and LEI, as compared to trailing edge strip i.e. TEO and TEI, to cater the turbulent loads produced by the airbrake.

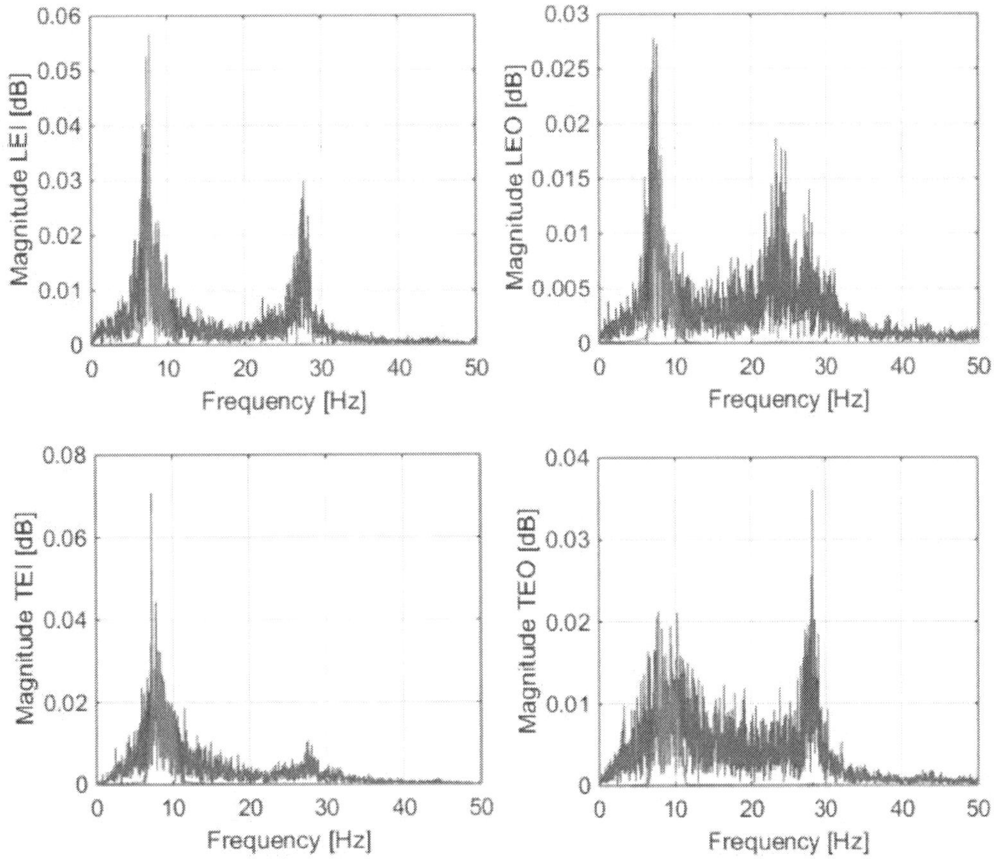

Figure 4. Actuator Movement during Bending Mode

Table 1. Hankel singular values and effective performance factor

Sensor Location / Control surface	Bending Mode HSV	Torsion Mode HSV	Bending Mode Effective Perf.	Torsion Mode Effective Perf.
LEI / LEI	×	0.118	4.711	1.665
LEO/ LEO	0.108	0.634	2.661	0.105
TEI/TEI	0.360	1.760	10.91	4.650
TEO/ TEO	2.032	×	14.94	×

Effective performance factor, **Error! Reference source not found.** can be extracted by normalizing the attenuation with the power consumption of each actuator during active control scheme.

Table 2. Effective performance of the sensors/actuators

Sensor Location / Control surface	Bending Mode	Torsion Mode
Midspan LEI / LEI	4.711	1.665
Wingtip LEO/ LEO	2.661	0.105
Midspan TEI/TEI	10.91	4.650
Wingtip TEO/ TEO	14.94	×

Conclusions
Aspect of suitable actuator/sensor configuration for modal attenuation is dealt in this work. Frequency response, power consumption, effectiveness and Hankel singular values are used experimentally to rank the performance of actuators in specific mode of interest. Trailing edge outboard outperformed other control surfaces for bending mode attenuation, while contrary to this result trailing edge inboard is the most suitable surface for torsion mode attenuation.

The presented research could pave the way for dedicated control surface selection for flight controls and flight dynamics and the add the redundancy based on optimized performance of the actuators and sensors.

References
[1] Patil, M. J., and Clark, R. L., "Aeroelastic Control Using Multiple Control Surfaces," 43rd AIAA/ASME/ASCE/AHS/ASC Structures, Structural Dynamics, and Materials Conference, Denver, Colorado, 2002.
[2] Pendleton, E., Bessette, D., Field, P., Miller, G., and Griffin, K., "Active Aeroelastic Wing Flight Research Program: Technical Program and Model Analytical Development," Journal of Aircraft, Vol. 37, No. 4, 2000, pp. 554–561.
[3] Perry, B., and Cole, S. R., "Summary of an Active Flexible Wing Program," Journal of Aircraft, Vol. 32, No. 1, 1995, pp. 10–15.
[4] Liebeck, R. H., "Design of the Blended Wing Body Subsonic Transport," Journal of Aircraft, Vol. 41, No. 1, 2004, pp. 10–25.
[5] De Gaspari, A., Ricci, S., Riccobene, L., and Scotti, A., "Active Aeroelastic Control Over a Multisurface Wing: Modeling and Wind-Tunnel Testing," AIAA Journal, Vol. 47, No. 9, 2009.
[6] Dowell, E. H., "A Modern Course in Aeroelasticity, Solid Mechanics and its Applications," 2015.
[7] Scotti, A., Ricci, S., and Quaranta, G., "Active control of a Three Surface Wind tunnel Aeroelastic Demonstrator: Modelling and Correlations," International Forum on Aeroelasticity and Structural Dynamics, Munich, Germany, 2005.

IOP Conf. Series: Materials Science and Engineering **899** (2020) 012013 doi:10.1088/1757-899X/899/1/012013

Stress Analysis of Composite Leaf Spring – Comparative Approach

Dr. Sheharyar Malik[1], Dr. Kamran S. Afaq[2]

[1] Instructor, Abu Dhabi Polytechnic, IAT, UAE.

[2] Head of Department Mechanical Engieering, HITEC University, Pakistan.

sheharyar.malik@adpoly.ac.ae

Abstract. Automobile manufacturers are focusing on weight reduction for saving natural resources and enhancing the fuel efficiency; automobiles are sharing almost half of the world's fuel consumption and a major source of pollution, thus reducing the weight will directly result in the reduction of fuel consumption. The leaf spring is one of the main candidates for the weight reduction in an automobile as they still holds the worth although many shocks absorbing devices are in the market these days. This paper is aimed to present the comparative study of steel and composite leaf spring with the later type having the same geometrical characteristics of steel leaf spring. Validations are carried out by modelling leaf spring on ANSYS & comparing it with analytical results. The results showed that the stresses in composite leaf spring are 64.8% lower than the stresses in the steel leaf spring for the same geometrical characteristics. The results for deflections showed also good agreement with analytical results and a decrement of 51.3% is found for composite leaf spring. The proposed methodology helps us to reduce the weight of leaf spring by approximately 74.39% without compromising on allowable stresses and thus also increasing the strain energy of the particular leaf spring.

1. Introduction

Vast and steep industrial growth causes great concern among the people for the consumption of natural resources. For this purpose scientists are trying to substitute new synthetic material for the structural parts that would replace the conventional material without any compromise on strength and performance. Composites are providing this breakthrough for researchers and although composites have replaced many conventional materials used for manufacturing, still great interest has been shown by automobile industry to use leaf spring made of composites due to their strength to weight ratio over conventional steel leaf springs. Leaf springs contribute around 10 to 20 % in the total body weight and the impact of weight reduction will be readily seen. Leaf springs can be primary candidates to be replaced by lighter materials like composites without any compromise on their load carrying ability [1], [2]. If the existing steel leaf springs are replaced with composite leaf springs, the impact in reduction of fuel consumption can be readily seen. Research activities devoted on the amount of vehicles in Pakistan and their fuel consumption found out that that 48 % of fuel is consumed by the transport sector [3], [4], [5].

Leaf springs are especially used due to their advantage over shocks and helical springs to carry heavy loads in terms of simplicity in maintenance and cost effectiveness and their significance is especially evident when heavy load systems are under consideration [6], [7]. Apart from acting as energy absorbing device, leaf springs also serves as a structural member whose ends could be guided along the pre-defined path, thus if placed properly as structural member they can also carry brake and driving torque in addition to providing cushion to impact and shocks loading [8].

Apart from reducing weight, leaf spring when manufactured from composites also offers increased strength to weight ratio and also it is noted that strain energy increases, because composites provides the necessary material properties having maximum strength and minimum modulus of elasticity in the fiber direction [9], [10]. Literature review helped to decide the glass epoxy as the most suitable combination for composite [11]. Literature also present the variants of conventional leaf springs, like mono steel leaf spring but it can prove catastrophic under fatigue or impact failure and it is the most prominent reason that they are not commonly utilized, weather its composite or steel leaf spring [12].

Semi elliptical multi leaf spring shown in Fig. 1, is selected for the analyses in this work. Moreover, over hanged condition is assumed to correctly model the forces at the center of the multi leaf spring. Both the eyes of the multi leaf spring are provided with the rotational degree of freedom, however one of the leaf spring eye is allowed to have the translation degree of freedom along the longitudinal direction.

In the presented work, comparative study has been carried out for the semi elliptical leaf spring, results are compared for the analytical and numerical analyses. The following section II presents the mathematical model for the design of the leaf spring. Section III will give insight about the numerical modelling of the leaf spring.

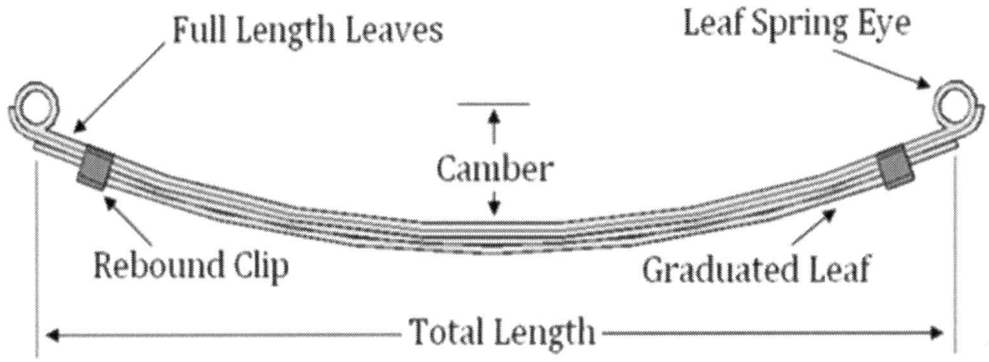

Figure 1. Semi Elliptical Leaf Spring.

2. Design of Leaf spring

2.1. Preliminary design

It is well known that the springs are designed to absorb and store energy and then release it. Hence, the strain energy of the material becomes a major factor in designing the springs. It can be easily observed that material having lower modulus and density will have a greater specific strain energy capacity. The introduction of composite materials made it possible to reduce the weight of the leaf spring without any reduction on load carrying capacity and stiffness. Since, the composite materials have more elastic strain energy storage capacity and high strength-to-weight ratio as compared to those of steel. The For simplicity leaf spring are idealized as cantilever beams. Mathematical modeling of leaf spring starts from simple assumptions and basic relation of bending stress and deflection for cantilever beam [3]:

$$\sigma = \frac{My}{I} \qquad (1)$$

Whereas, 'M' is bending moment, 'y' is the distance from neutral axis to surface and 'I' is moment of inertia.

2.2. Mathematical design

A little consideration will show that the spring become compact so that the space occupied by the graduated springs is considerably reduced. The graduated leaves may have zero width at the loaded end. But sufficient metal must be provided to support the shear stress. Therefore it becomes necessary to have one or more leaves of uniform cross section extended clear to end. Since the master leaf has to with the stand vertical bending load as well as load due to sideway of a vehicle and twisting it is usually to provide at least two full length leaves and the rest graduated leaves. The equations of bending stress and deflection can be rewritten as a special case for different number of full length leaves σ_F and graduated leaves σ_G:

$$\sigma_F = \frac{18WL}{bt^2(2n_G + 3n_F)} \quad (2)$$

$$\sigma_G = \frac{12WL}{bt^2(2n_G + 3n_F)} \quad (3)$$

$$\delta = \frac{12WL^3}{Ebt^3(2n_G + 3n_F)} \quad (4)$$

Where, 'W' represents the force on the eye of the leaf spring, 'L' represents the total length, 'E' represents the modulus of elasticity, 'b' represents the breadth of the leaf spring cross section, 't' represents the thickness of the leaf spring cross section. Number of graduated leaves and full length leaves are represented by n_G and n_F, respectively.

3. Numerical Modelling

While designing the composite leaf spring, several design criteria are considered. In the presented work, similar geometrical properties are chosen and the difference in the stresses is noted for the steel and composite leaf spring. Commercially available software, ANSYS is used to design the model and to carry out the analysis on leaf spring. Element type, SOLID 45 is selected for the analysis. The movement of the graduated leaves in the transverse direction is limited by constraining the leaves in the transverse direction and allowing a slip in the longitudinal direction.

3.1. Contact definition

It is obvious from Fig. 1, the leaves of the spring are in contact with each other while allowing a slip motion along the longitudinal direction. This calls for the definition of appropriate surface to surface contact mechanism where it permits the motion of the leaves along the longitudinal direction. In order to examine the correct surface to surface contact the supplementary analysis is carried out, in which two beams are modeled as shown in Fig. 2, validations are carried out for different contact types by comparing it with the analytical results.

Nodes are selected to make the flexible sliding contact between two surfaces of two leaves with the help of CONTACT 174 and TARGET 170 elements. Several contact types definitions are available in the literature, most prominent are rough, bonded and no separation contact type. Rough contact type models perfectly the rough frictional contact where there is no sliding and this corresponds to an infinite friction coefficient. Bonded, models a contact in which the surfaces are bonded in all directions. For the presented case, 'no separation' is found to be the most suitable contact type which models a contact in which the surfaces are tied (although sliding is permitted) with the value of normal penalty stiffness as 1.0 and penetration tolerance is chosen as 0.1. Path plots were also plotted to infer the behavior of stresses within the leaves in Fig. 3. Comparison for no separation, rough and bonded contact type are presented in Table 1.

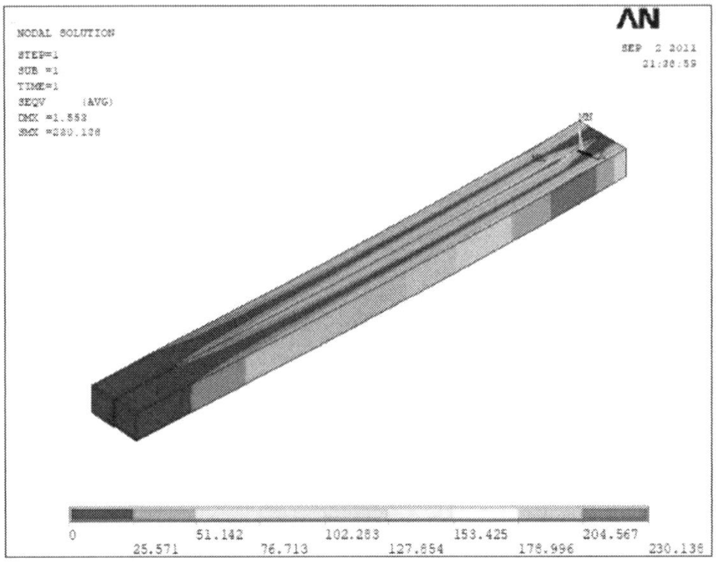

Figure 2. Numerical analysis - Contact type - No Separation – Load 100N

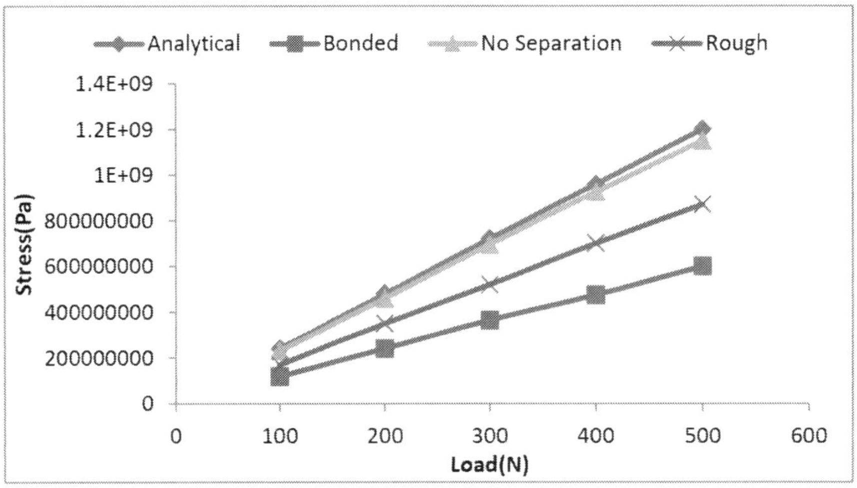

Figure 3. Load vs. Stress - Contact Types

Table 1. Comparison of surface contact

	Analytical		Bonded Contact		Error Bonded		No Separation Contact		Error No Sep.		Rough Contact		Error Rough	
Load	σ	δ	σ	δ	σ	δ	σ	δ	σ	δ	σ	Δ	σ	δ
500	1200	7.92	600	1.946	50	75.4	115	7.771	4.17	1.881	870	3.326	27.5	58.0

3.2. Modelling of composite

The leaf is designed to provide the strength in the direction along which major portion of load is distributed i.e. we can direct fibre orientation in the direction of load. Glass/Epoxy in the direction of fibres has good characteristics for storing strain energy. Table 2, presents the comparison of the steel and Glass/Epoxy composite.

Table 2. Comparison of material properties

Material	σ/ρ	E/ρ
Steel	2.7	2.7
Glass/Epoxy	13.6	2.9

The unidirectional lay up along the longitudinal direction of the spring is selected with a minimum of 10 % layers in each direction as shown in Table 3 and depicted in Fig. 4. Moreover, maximum fibres are orientated along the longitudinal direction. Thickness of each layer is also important part of design. Total height was 26mm for leaf spring, so it was divided into 100 layers of thickness 0.26 mm in the z-direction. The density and weight of weight of composite leaf spring for fibre fraction of 60% can be evaluated as follows:

$$\rho_c = V_f\rho_f + V_m\rho_m \tag{5}$$

Whereas, $V_f = 0.6$,

$V_m = 1 - V_f$,

$\rho_f = 2540\ Kg/m^3$, $\rho_m = 1100\ Kg/m^3$ is selected for Glass/Epoxy.

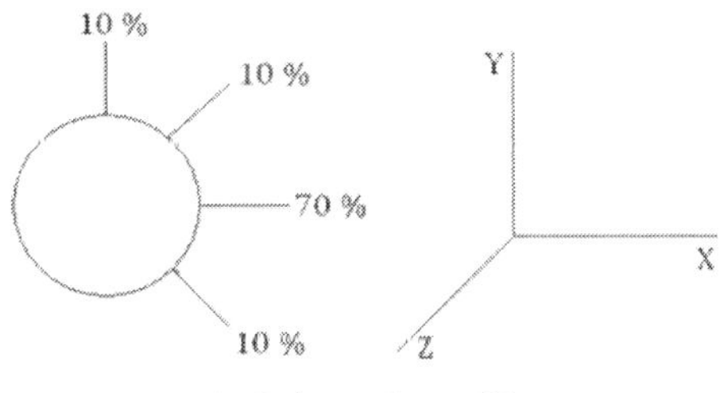

Figure 4. Orientation of Layers

Table 3 shows the results for composite leaf spring.

Table 3. Comparison of stress for different layer combinations

Glass Epoxy Combinations	Stress(σ) MPa
[(0/0/45/0/0/-45/0/0/90/0)$_5$]$_S$	85.82
[(90/90/45/90/90/-45/90/90/0/90)$_5$]$_S$	150.29
[(45/90/0/0/-45/45/90/90/0/-45)$_5$]$_S$	140.82

IOP Conf. Series: Materials Science and Engineering **899** (2020) 012013 doi:10.1088/1757-899X/899/1/012013

Comparison is also carried out for the changed geometry i.e. comprising of two leaves and mono leaf spring. Multi leaf spring analysis is preferred due to catastrophic fatigue failure of leaf spring comprising of single/mono leaf [4].

Weight reduction of 74.39 % is achieved when steel leaf spring is replaced with composite leaf spring as shown in Fig. 1b., while Fig. 1a. shows one of the analysed case. Fig. 2a. and Fig. 2b. shows the deflection and stress along the main leaf spring, respectively. Weight reduction of 74.39 % is achieved when steel leaf spring is replaced with composite leaf spring. Several combinations for layer orientations were analysed. It was found out that minimum stress in leaf spring were when maximum layers were placed in the longitudinal direction. It is further illustrated with the help of figures and Table 4 presents the comparison for the steel and composite leaf spring.

Table 4. Comparison of steel and composite leaf spring

	Composite Leaf Spring		Steel leaf spring		Difference	
Load	σ	δ	σ	δ	σ	δ
N	Pa	mm	Pa	mm	%	%
500	4.29E+07	5.839	1.26E+08	11.492	65.8	49.19074

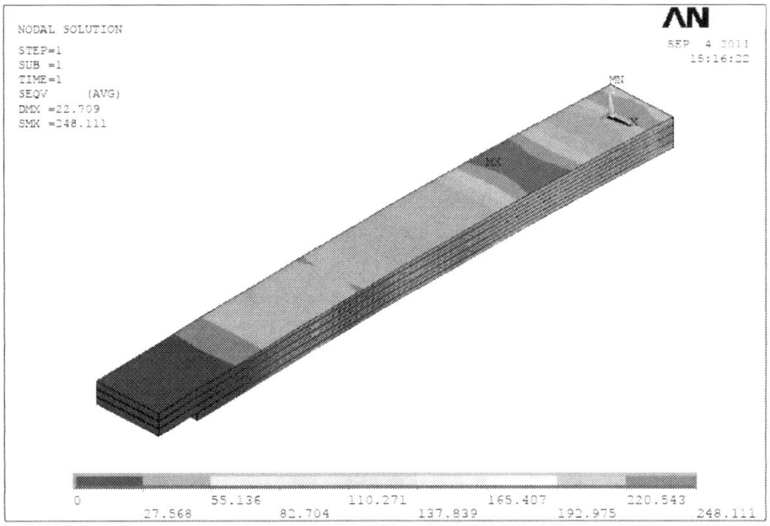

Figure 5. Stess Analysis of composite leaf spring

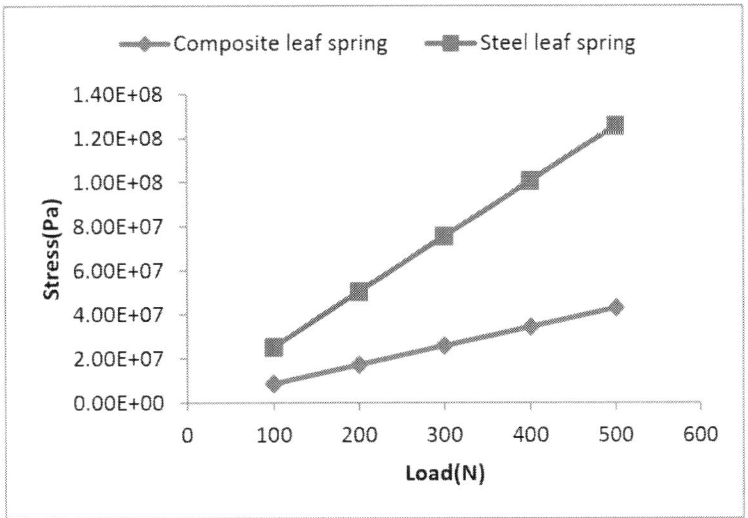

Figure 6. Stess Analysis of composite leaf spring

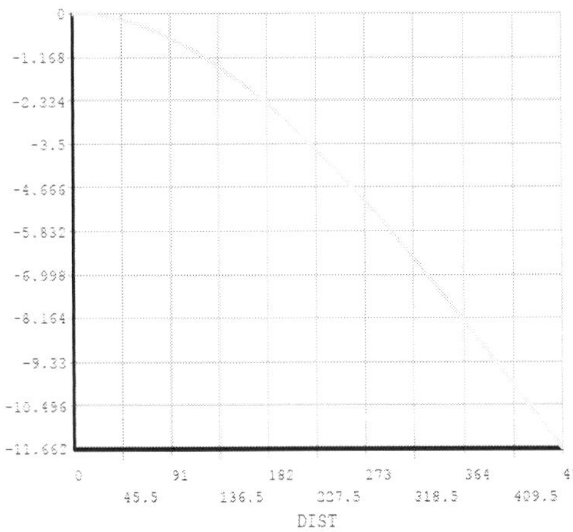

Figure 7a. Length (mm) vs. Deflection (mm)

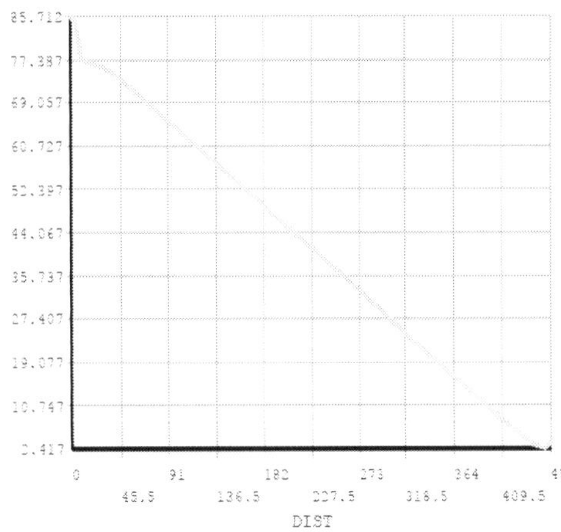

Figure 7b. Length (mm) vs. Stress (MPa)

4. Conclusion

The results showed that the stresses in composite leaf spring are 64.7% lower than the stresses in the steel leaf spring of same dimensions. The results for deflections showed also good agreement with theoretical results and when compared with steel leaf spring a decrement of 51.3% is found for deflection in the composite leaf spring. This methodology helps us to reduce the weight of weight of leaf spring by approximately 74.39 % without compromising on allowable stresses and thus also increasing strain energy for our particular leaf spring case. Numerical analysis is used to illustrate the stresses along the leaves in full length and graduated length leaves, it is found to be corresponding to the theory for the beams that stress in the graduated leaf should be nearly 2/3 of the stress in full length leaf.

Leaf spring is used in almost every heavy vehicle used for transportation of goods and passengers. In Pakistan leaf spring can also be found in light vehicles such as Suzuki Mehran, Bolan and Ravi (Pickup). By reducing the weight of existing steel leaf spring will help to reduce the weight of these vehicles by 10 – 20 % percent, which will help in reducing the fuel consumption up to nearly 6 %.

IOP Conf. Series: Materials Science and Engineering **899** (2020) 012013 doi:10.1088/1757-899X/899/1/012013

References

[1] P. Breadmore, C. F. Johnson, "The Potential for Composites in Structural Automotive Applications", Composite Science and Technology 26, 2006, pp 251-281.

[2] R. L. Daugherty, "Composite Leaf Springs in Heavy Truck Applications", International Conference on Composite Materials, Proceeding of Japan – US Conference, Tokyo 2001, pp 529-538.

[3] Trading Economics, "Road Sector Gasoline Fuel Consumption in Pakistan", World Bank Indicators – Pakistan – Transportation, 2011.

[4] Earth trends, "Energy and Resources - Pakistan", Earth trends, 2009.

[5] A. Ahmed, M. K. Jha, "Status of Petroleum Sector in Pakistan", Oil and Gas Business, Sian Institute of Technology, Bangkok, Thailand, 2008.1989.

[6] J. K. Hedrick, K. Yi, "The Effect of Alternative Heavy Truck Suspensions on Flexible Pavement Response", Working Paper, No. 46, September 1991.

[7] N. Philipson, Modelon AB, "Leaf Spring Modeling", Ideon Science Park SE -22370 Lund, Sweden.

[8] M. H. M. Goudarzi, A. Masoumi, "The Effects of Replacement of Steel Leaf Spring with a Composite One on Different Aspects of Vehicle Dynamics", International Conferences on Advances in Mechanical Engineering, AME, 2006.

[9] R.S. Khurmi, "A Textbook of Machine Design", Chapter 23, Springs, pp 866-869.

[10] M. S. Kumar, S. Vijayarangan, "Analytical and Experimental Studies on Fatigue Life Prediction of Steel and Composite Multi Leaf Spring for Light Passenger Vehicles Using Life Data Analysis", ISSN 1392-1320, Materials Science, Vol. 13, No. 2, 2007.

[11] "Spring Design Manual", Design and Application of Leaf Springs, AE-11, Society of automotive Engineer HS-788, 1990.

[12] G. S. Shiva Shankar, S. Vijayarangan, "Mono Composite Leaf Springs for Light Weight Vehicles – Design, End Joint Analysis and Testing", ISSN 1392-1320, Materials Science, Vol. 12, No. 3, 2006.

IOP Conf. Series: Materials Science and Engineering **899** (2020) 012014 doi:10.1088/1757-899X/899/1/012014

Predicting compressive strength of CRM samples using Image processing and ANN

M I Waris[1], J Mir[2], V Plevris[3] and A Ahmad[1, 4]

[1] Department of Civil Engineering, University of Engineering and Technology Taxila, Punjab, Pakistan
[2] Department of Electrical Engineering, University of Engineering and Technology Taxila, Punjab, Pakistan
[3] Department of Civil Engineering and Energy Technology, OsloMet- Oslo Metropolitan University, Oslo, Norway

[4] Corresponding author E-mail: afaq.ahmad@uettaxila.edu.pk

Abstract. Quality of concrete is majorly ascertained through its compressive strength which has a significant role in the stability of concrete structures. In this study, artificial neural network (ANN) and image processing (IP) techniques were used to predict the concrete compressive strength (f_c) with cement replacement material (CRM), i.e., Fly Ash (FA) and Silica Fumes (SF). 18 concrete cylinders were cast with different mix ratios and with different % of CRM. Half of them were tested for compression strength in the laboratory and remaining cylinders were cut into three slices each, for prediction of compressive strength through the proposed technique. Images were obtained using a DSLR camera under defined conditions to extract the features. Based on the extracted features, ANN modelling was performed for predicting f_c. A comparison of experimental results and ANN results (R = 0.9865) proved ANN models can be used as a prediction tool for compressive strength of concrete.

Keywords—Artificial neural network, image processing, concrete, compressive strength,

1. INTRODUCTION

Machines are employed to devise algorithms from the acquired data to predict desired outcomes. Artificial neural networks (ANN) and image processing (IP) are popular techniques in artificial intelligence utilized to solve complex and time-consuming problems in the field of engineering and science. ANN is a computational network consisting of different interconnecting layers and neurons used to perform the duties of a biological mind. IP is the process of extracting useful information from images with the aid of a computer algorithm. From the past few decades, researchers have been using ANN and IP in the field of civil engineering to study building materials, concrete technology, structural engineering, and morphology of building materials [1-9]. Amongst the building materials, concrete: a composite material made up of cement, sand, and aggregate bonded together [1, 10], has a major role in the stability of structures. Concrete exhibits heterogeneous structure with a non-linear behavior and is extensively used in buildings, bridges, dams, and roads because of its nature of high compressive resistance. The main characteristic of concrete is the compressive strength f_c , which has a direct impact on other characteristics that can be measured from the concrete compressive strength [11].

Based on the application, the compressive strength f_c of concrete can be studied through various conventional test methods, such as Schmidt hammer, ultrasonic pulse velocity, core drilling, electrical resistivity measurement, etc. These methods can be mainly categorized into two approaches: destructive and non-destructive testing [5, 12]. Though efficient, these methods demand specialized equipment for making concrete compressive strength measurements and are, therefore, labor-intensive, expensive and time-consuming. To address these shortcomings, a new non-destructive technique is recently proposed

IOP Conf. Series: Materials Science and Engineering **899** (2020) 012014 doi:10.1088/1757-899X/899/1/012014

in [11] as an alternative to traditional concrete compressive strength measuring methods. The method estimates the compressive strength of concrete with an accuracy of 99.8% using ANN and IP. However, the effectiveness and validation of this vision-based approach is not investigated when cement replacement material is utilized partially in the concrete.

In this paper, a computer-vision based approach employing ANN and IP is presented to predict the compressive strength f_c of the concrete. For this, 18 cylindrical samples with different mix design ratios and cement replacement material percentages were cast and cured. Then, concrete surface images were acquired from which statistical features such as mean, median and standard deviation were extracted using IP. These features were utilized for training the ANN architecture to predict the compressive strength which is then compared with the actual measurements for validation.

The rest of the paper is organized as follows. Section 2 provides the specimen preparation and testing details. Section 3 discusses the IP algorithms and ANN architecture, utilized for image acquisition, feature extraction, and training. Finally, Section 4 concludes the paper with results and discussions.

2. SPECIMEN PREPARATION AND TESTING

2.1. Materials and Mixing

For casting the concrete samples, ordinary portland cement (OPC) as per the specification of ASTM C150 Type I with aggregate and river sand (locally available in Pakistan) were used. Cement replacement materials (CRM) such as Silica fume (SF) and fly ash (FA) were utilized to replace cement by volume in the mixture by 0%, 15%, 25%. The chemical compositions of OPC, SF and FA are given Table 1. Selected mix proportions to prepare different concrete mixes are tabulated in Table 2. Mixtures were prepared under the natural environment. Then, 18 concrete cylindrical specimens of dimension 150 mm × 300 mm were cast and cured in the natural water for 14 days. The concrete samples were left in the air to dry for 24 hours before performing compressive strength tests.

Table 1. Chemical composition of cement, silica fume and fly ash.

	Cement	SF	FA
Silica	22.5	84 − 86	57 − 65
Aluminum oxide	5.0	1.0 (max.)	28 − 32
Iron oxide	4.0	2.0 − 3.5	1 − 4
Calcium oxide	64.25	1.0 − 1.5	1 − 2
Loss on ignition	0.64	4 − 7	9.01

Table 2. Samples mix ratio with CRM by percentage of volume of cement.

Sample	Mix Ratio	W/C Ratio	% FA	% SF	14 Days Curing
7S0.4W0A	1:2:4	0.4	0	0	2
7S0.5W0A	1:2:4	0.5	0	0	2
7S0.6W0A	1:2:4	0.6	0	0	2
7S0.4W15A	1:2:4	0.4	15	15	2
7S0.5W15A	1:2:4	0.5	15	15	2
7S0.6W15A	1:2:4	0.6	15	15	2
7S0.4W25A	1:2:4	0.4	25	25	2
7S0.5W25A	1:2:4	0.5	25	25	2
7S0.6W25A	1:2:4	0.6	25	25	2

2.2. Compressive Strength f_c

The casted cylindrical concrete samples containing cement replacement material *FA* and *SF* with 0%, 15%, 25% by weight of cement were tested to obtain the compressive strength f_c . The f_c was measured according to ASTM 39 on a cylindrical specimen of dimension 150 mm × 300 mm for each mix ratio cured at 14 days.

3. IMAGE PROCESSING AND ARTIFICIAL NEURAL NETWORK

3.1. Image acquisition setup

As the efficiency of any image processing based technique largely depends upon the quality of the images, special arrangements were made to ensure that the acquired images of the prepared cylindrical specimens were free of any noise, light variation or shadows. For this, the imaging lab was built in a small room of dimension 3 m × 3 m. First, the sunlight was blocked by using black sheets on each entry point of light. Then, two 30W LED bulbs were used to produce artificial light of 2000 lux to illuminate the specimens [11] for image acquisition. This ensured the uniform illumination in the room which allowed to capture images without generating any unwanted shadows.

3.2. Cutting of concrete samples

Nine of the 18 concrete cylinder samples were cut horizontally using sawing machine into three slices with each slice having the same size (150 mm diameter and 101.6 mm height). Top and bottom faces of the cylinder slices were labelled, as depicted in Fig.1.

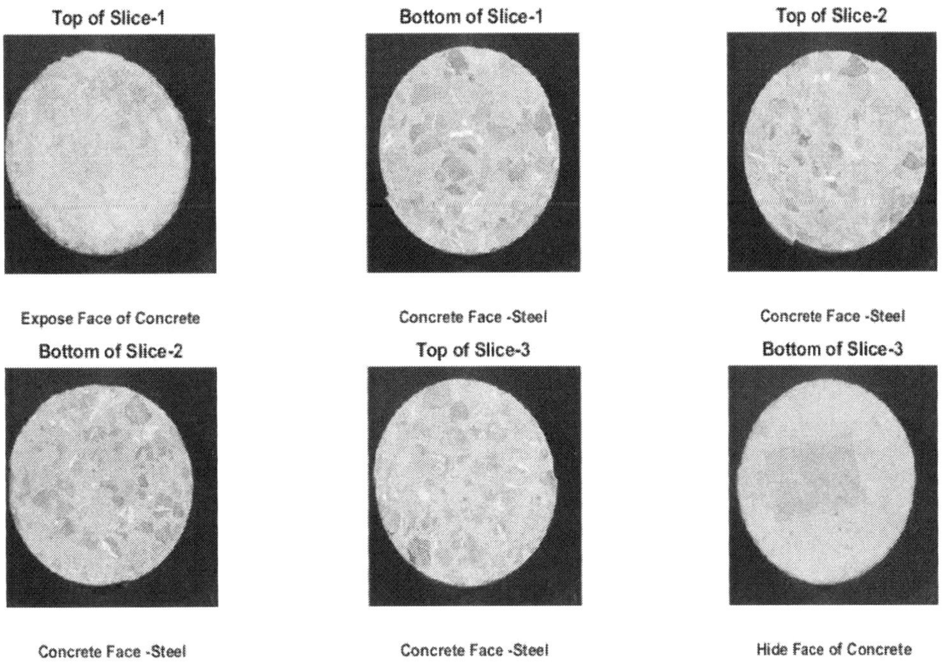

Figure 1. Top and bottom faces of three slices of concrete cylinder.

3.3. Image acquisition and Feature Extraction

Pictures of the slice surfaces of the specimen were acquired with DSLR NIKON 3300 camera mounted 300 mm vertically above the sample surface. The camera had a 24-million-pixel capacity. Accounting the top and bottom faces of three slices, six colored images in total were taken for each concrete cylindrical specimen. The captured images are digital where each value, generally referred to as a pixel, represents the brightness of the image at that point. Then, the images were converted into the grayscale representation I using (1), where R, G and B represents the red, green and blue colour pixel values, respectively.

$$I = 0.2290 \times R + 0.558 \times G + 0.113 \times B \quad (1)$$

After the grayscale conversion, the images are resized to 256 × 256 resolution for further processing. To ensure the structural details are not changed due to down-sampling, the image is low pass filtered before the application of resizing. Once the images are resized, statistical features like mean $\overline{I}$, median $\tilde{I}$, and standard deviation σ_I^2 are to be calculated using (2), (3) and (4), respectively [10], where n is the total number of pixel values in grayscale image I.

$$\overline{I} = \frac{1}{n}\sum_{i=1}^{n} I \quad (2)$$

$$\tilde{I} = \{(n+1) \div 2\}^{Th} \quad (3)$$

$$\sigma_I^2 = \sqrt{\frac{1}{n}\sum_{i=1}^{n}(I-\overline{I})^2} \quad (4)$$

The statistical features are to be extracted from region of interest only, i.e., the concrete surface in the image and not from the background portion of the image. To ensure this, the masks are generated as shown in Fig. 3 using thresholding technique by extracting the threshold value from the histograms depicted in Fig. 2.

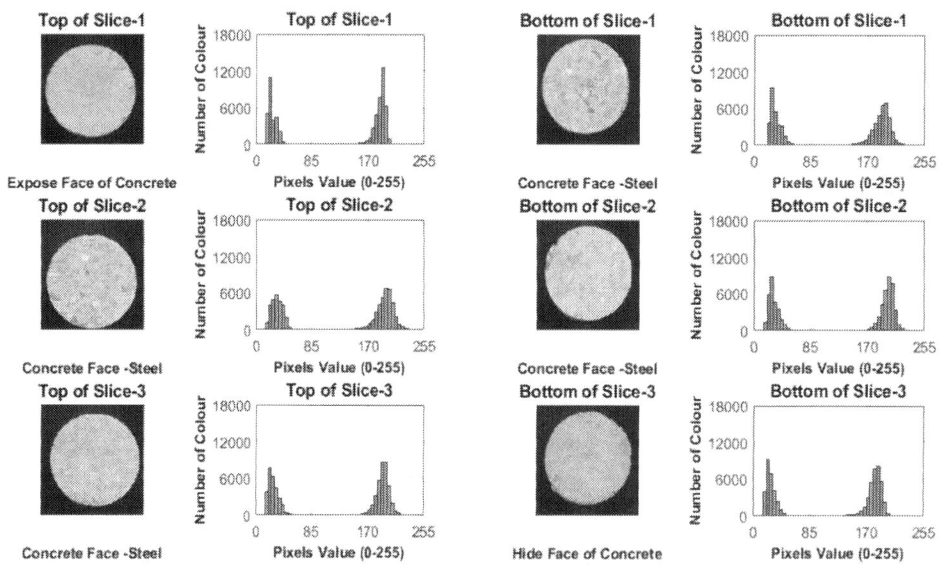

Figure 2. Histogram of pixel values of six faces of concrete cylinder slices.

IOP Conf. Series: Materials Science and Engineering **899** (2020) 012014 doi:10.1088/1757-899X/899/1/012014

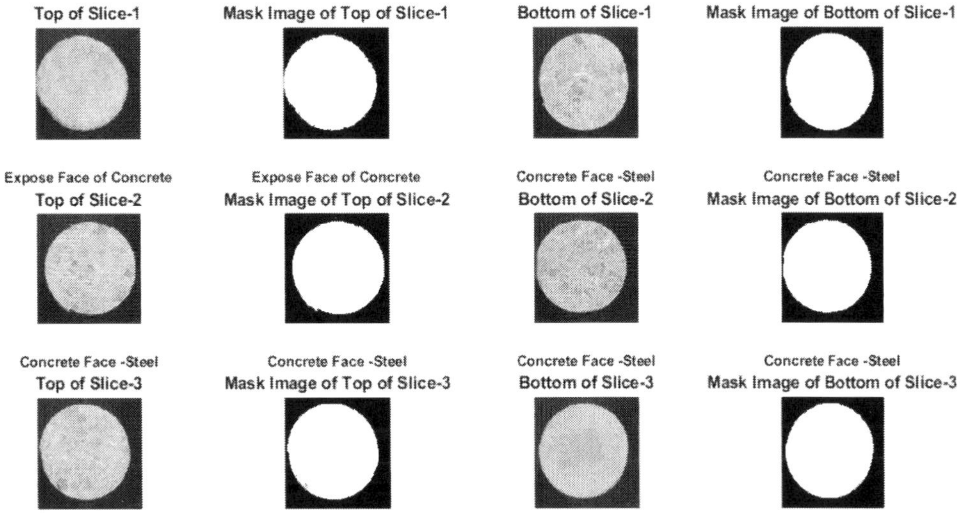

Figure 3. Masking of top and bottom faces of three slices of concrete cylinder.

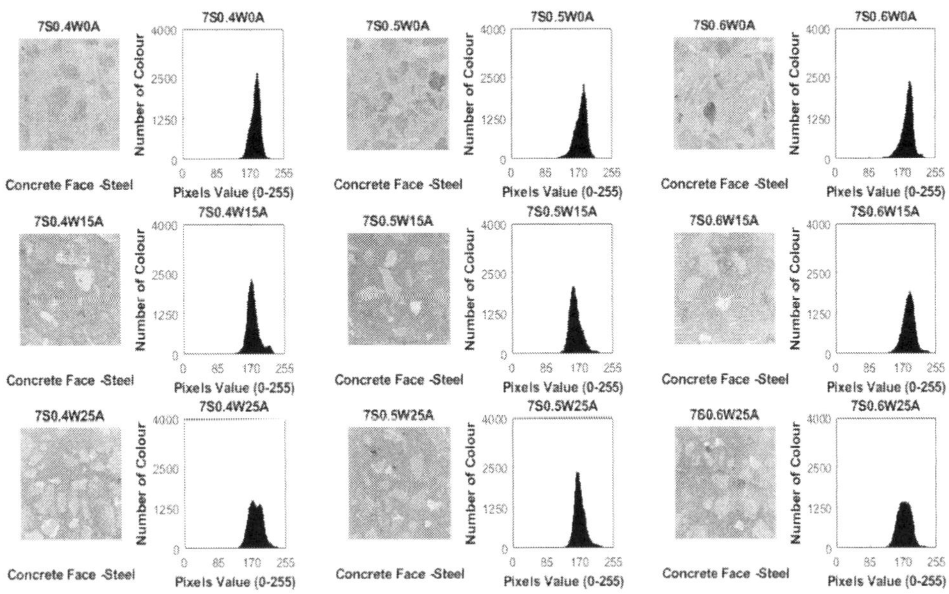

Figure 4. Histogram of center faces of concrete specimen of all mix ratios.

The effect of *SF*, *FA* and water cement ratio (W/C) on air voids can be observed in Fig. 4, which shows the top concrete image of the centre slice for all mix ratios after the application of masking. Clearly, the distribution of histograms is different for different mix ratios from which it can be inferred that statistical features will characterize the image representation and thereby its properties like compressive strength.

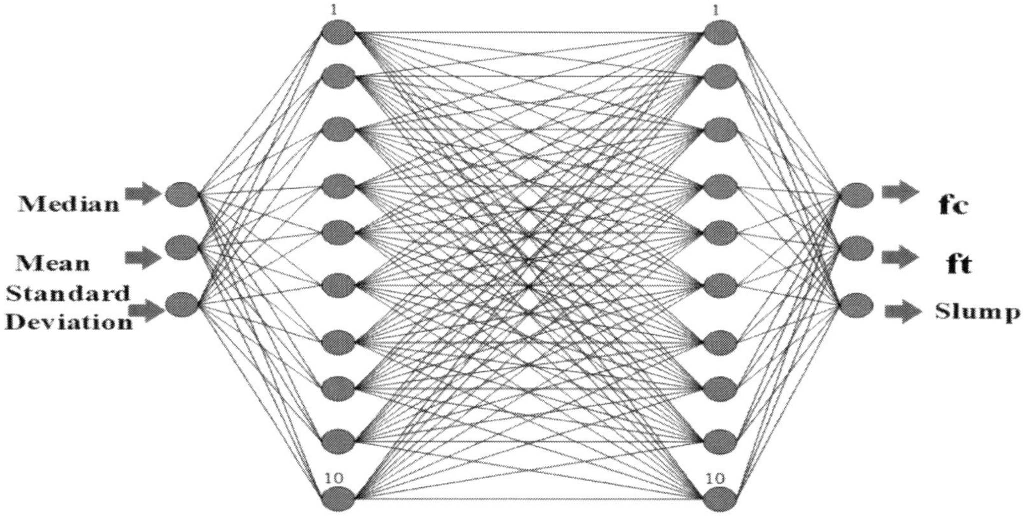

Figure 5. Architecture of the optimized ANN model.

3.4. ANN architecture and Training

ANN is a set of layers and nodes used to process information similar to the human brain. In our method, the Levenberg-Marquardt (LM) back-propagation ANN model was adopted to process the data. The extracted statistical features of standard deviation, mean and median were taken as input and experimental values of compressive strength, tensile strength, and slump obtained from laboratory tests were used as target values. Before training, normalization of input and output values between 0 and 1 was done using (5), where Z_s represents normalized value of the variable u and u_{min} and u_{max} are the minimum and maximum values of the variable u, respectively.

$$Z_s = \frac{(u - u_{min})}{(u_{max} - u_{min})} \qquad (5)$$

The optimum model was found after many trails and adjusting learning values and moment constant [11]. The architecture of the optimized ANN model with one input layer, two hidden layers with ten neurons each and one output layer for this study, is shown in Fig. 5. After developing the initial ANN architecture, the training process was repeated iteratively to obtain an optimum ANN model. An optimized model was obtained with 10000 iteration number which was tested against various tuning parameter values during the process. The model that gives the highest accuracy was selected. The selected iteration value is the ideal iteration number related to the sampled problem. So, the best ANN model was found for the minimum training and testing error. The formula for the training and testing errors is given in (6)

$$ET(\%) = \left(\sum_{n=1}^{k} \frac{|a(n) - b(n)|}{m * j} \right) * 100 \qquad (6)$$

Here, $a(n)$ represents the desired output values, $b(n)$ represents the ANN outputs, k represents the number of data points in the training and testing data, m represents segment numbers in the training and testing data, and j represents the ANN outputs for the training and testing operations.

4. RESULTS AND DISCUSSIONS

In the presented work, a combination of two artificial techniques IP and ANN was applied to determine the compressive strength of concrete, which is the significant construction material in civil engineering. Experimental results and the results from the adopted method were compared and found similar. The accuracy of the results obtained by our technique is 98.65%. Following are few key observations.

- Numerous factors affect the mechanical properties of concrete. A data set could be produced including all such parameters water/cement ratio, curing, additives, the quantity of cement, compaction, size of aggregate, size and shape of the specimen.
- In this study, the ANN model with two hidden layers and 10 nodes was selected, other effective parameters lr and mc were 0.9 and 0.2, respectively. Chosen parameters affect the performance of ANN such as processing time, accuracy, etc.
- In the present study, different parameters are chosen for concrete are for a different purpose. Different combinations of parameters are used mainly because of their effect on the mechanical properties of concrete. Therefore, results are specific to the parameters used which is a limitation of the current study.

5. REFERENCES

[1] A. J. Tenza-Abril, Y. Villacampa, A. M. Solak, and F. Baeza-Brotons, "Prediction and sensitivity analysis of compressive strength in segregated lightweight concrete based on artificial neural network using ultrasonic pulse velocity," *Construction and Building Materials,* vol. 189, pp. 1173-1183, 2018.

[2] C.-Y. Kao, C.-H. Shen, J.-C. Jan, and S.-L. Hung, "A Computer-Aided Approach to Pozzolanic Concrete Mix Design," *Advances in Civil Engineering,* vol. 2018, pp. 1-15, 2018.

[3] T. Gupta, K. A. Patel, S. Siddique, R. K. Sharma, and S. Chaudhary, "Prediction of mechanical properties of rubberised concrete exposed to elevated temperature using ANN," *Measurement,* vol. 147, 2019.

[4] P. Chopra, R. K. Sharma, and M. Kumar, "Prediction of Compressive Strength of Concrete Using Artificial Neural Network and Genetic Programming %J Advances in Materials Science and Engineering," vol. 2016, p. 10, 2016, Art. no. 7648467.

[5] J. A. Bogas, M. G. Gomes, and A. Gomes, "Compressive strength evaluation of structural lightweight concrete by non-destructive ultrasonic pulse velocity method," *Ultrasonics,* vol. 53, no. 5, pp. 962-972, 2013/07/01/ 2013.

[6] S. Petruseva, V. Zileska-Pancovska, and D. Car-Pušić, "Implementation of Process-Based and Data-Driven Models for Early Prediction of Construction Time," *Advances in Civil Engineering,* vol. 2019, pp. 1-12, 2019.

[7] A. Madadi, H. Eskandari-Naddaf, R. Shadnia, and L. Zhang, "Characterization of ferrocement slab panels containing lightweight expanded clay aggregate using digital image correlation technique," *Construction and Building Materials,* vol. 180, pp. 464-476, 2018.

[8] N.-D. Hoang, "An Artificial Intelligence Method for Asphalt Pavement Pothole Detection Using Least Squares Support Vector Machine and Neural Network with Steerable Filter-Based Feature Extraction," *Advances in Civil Engineering,* vol. 2018, pp. 1-12, 2018.

[9] T. Dede, M. Kankal, A. R. Vosoughi, M. Grzywiński, and M. Kripka, "Artificial Intelligence Applications in Civil Engineering," *Advances in Civil Engineering,* vol. 2019, pp. 1-3, 2019.

[10] G. Doğan, M. H. Arslan, and M. Ceylan, "Statistical Feature Extraction Based on an Ann Approach for Estimating the Compressive Strength of Concrete," *Neural Network World,* vol. 25, no. 3, pp. 301-318, 2015.

[11] G. Dogan, M. H. Arslan, and M. Ceylan, "Concrete compressive strength detection using image processing based new test method," *Measurement,* vol. 109, pp. 137-148, 2017.

[12] D. Breysse, "Nondestructive evaluation of concrete strength: An historical review and a new perspective by combining NDT methods," *Construction and Building Materials,* vol. 33, pp. 139-163, 2012/08/01/ 2012.

Experimental study of railway bridge scouring and proposal of remedial measures-A case study of bridge number 30, Harrow River.

Muhammad Sohail Javed[1,3,a], Naeem Ejaz[2,b]

[1] Civil Engineering Department, University of Engineering and Technology, Taxila, Pakistan.
[2]Civil Engineering Department, University of Engineering and Technology, Taxila, Pakistan.
[3]Pakistan Railways, Rawalpindi Divison, Pakistan.

[a)]engineermuhammad88@gmail.com , [b)]naeem.ejaz@uettaxila.edu.pk

Abstract. Bridges are one of the most integral transportation components for connection of remote parts worldwide. The concern of every engineer is with their stability and collapse. One of the issues with bridge stability is bridge pier scouring in perennial and inundation canals as well as rivers, If not properly monitored and precautionary measures not adopted, it can result into bridge collapse. This paper makes an attempt to study the heavy bridge scouring in Railway Bridge (bridge # 30) on Harrow River located at downstream of Khanpur Dam, Pakistan. The study is carried out experimentally on scaled down model of bridge to study the behavior and patterns of scouring in the river bed during and after floods in two phases including 1) investigating the causes of scouring 2) provision and analysis of suitable remedial measures. The scouring that occurred in lab model matched the field patterns. Baffle walls and weir were provided separately as remedial measure but the former failed under initial watering while weir provided on downstream side reduced scouring by 95%.

Keywords—bridge, scouring, experimental model, remedial measures, baffle walls, weir

1. Introduction

Scouring is a natural phenomenon of water carrying away sediments from river bed and riverbanks, around the bridge piers and abutments caused by erosive action of the flowing water of alluvial channels. The construction of bridges in alluvial channels causes a contraction in the waterway at the bridge site. Resultantly, this contraction in the waterway causes a significant scour at that site. Many bridges failed around the world because of extreme scour around piers. The vertical reduction in the river bed measured from bed level is called scour depth (Thamer et al. 2005). The longitudinal and transverse extents of scour increase with scour depth. The volume of scour hole increases quadratically with maximum scour depth after 24 h. (Rashid Farooq, august 2019). Studies reveal that maximum scour takes place on the upstream and side of the pier. Accordingly, efficacy of scour protection measures such as raft foundation, cut-off walls can be studied and verified. (Bruce .w.melville). For live-bed scour, the existence of a downstream submerged weir can cause upstream bed aggradations, reducing the scour depth at the pier (. Lu Wang, 2018)

2. Objectives:

The aim of this research is to experimentally stimulate bridge scour conditions of Railway bridge number 30 on Harrow River, Pakistan. After stimulation the causes are to be investigated and then suitable remedial measures to be provided and their efficacy experimentally checked. Finally, scour prevention measures based on experimental verification to be proposed for the respective bridge in field.

3. Experimentation:

3.1. Case study description:

Pakistan Railways is one of the biggest state owned organizations. Bridges having single clear span more than 60 ft are called major bridges in Pakistan Railways. Bridge #30 is a major (12 spans of 65 feet) and one of the most important bridges for CPEC project. Being on downstream of Khanpur dam there is always a threat of flood situation. B # 30 was completely washed out in year 1997 floods and then it was reconstructed in year 2000. But again in 2007, due to heavy flood in Khanpur dam, the high water velocity eroded the bridge bed and 30 feet deep channel was formed along with 20' piles which were exposed to environment. For temporary cater of scouring 13000ft3 stones were laid around bridge but it did not prove to be of any great significance. The bridge sections are shown in figure 1 and 2.

Figure 1. Railway Bridge #30

Figure 2. Scoured bridge piers

3.2. Experimental set up of model and river bed:

This study was carried out in a flume of fixed slope with a length of 10 m and a width of 0.96m at Hydraulic Laboratory of University of Engineering And Technology, Taxila. Depth, flow and velocity of water was controlled both by valve of inlet pipe at upstream of flume and controlled gate at the end of downstream of flume. Velocity was measured with electronic current velocity meter and discharge was measured with the help of triangular weir at the end of flume and also by equation

$$Q=AV$$

Here q is discharge in meter/ cubic sec, A is cross sectional area, V is velocity in metre/ sec. Scale down model of 1:50 of bridge # 30 was built, 2 piers with 4 piles (0.393cm) each and with exact same river bed material (type) in flume. Experiments were performed on one of the four effected spans out of twelve spans of b#30. The bridge setup in flume was built at a distance of 3m from water inlet so that turbulent effect due to entering water does not affect actual study area (turbulence affect is generally up to 2m), this also resulted in uniform water flow and water profile. River bed material was filled in flume up to height of 0.2m and up to length of 15m and at the end of downstream soil material stone blocks were placed to retain material. The setup is shown in figure 3.

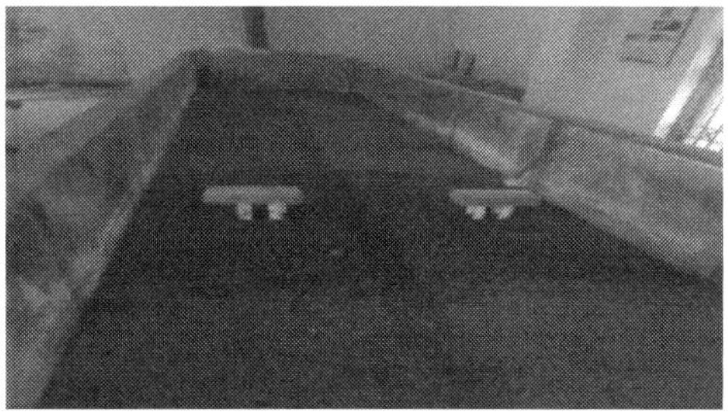

Figure 3 Bridge model in flume

Figure 4 Weir model on downstream of bridge

3.3. Test:

In each experiment water was allowed to run for 24hrs and at the end of experiment water was drained off very carefully so that flume bed couldn't be disturbed. Scouring was measured both during live water and at the end of complete water drainage with point gauge (with +-0.5mm error). Experiments were conducted under clear water conditions.

3.4. Remedial measure set-up:

A weir (fig 4) and baffle walls were also provided separately at downstream of piles to study as remedial measures. Weir was provided at a distance of 2m from piles while 3 baffle walls were provided at a distance of 0.30m from piles and with spacing of 0.30m.

4. Results and Discussion:

The following are the results observed in the established laboratory model of the bridge number 30;

- After conducting a series of experiments in hydraulic flume at different discharges from as low as 0.03675cumecs to as high as .060192cumecs, it was observed that live water scouring occurs at the very early stage and at very high discharges i.e. 0.03675Cumecs and above. Clear water stage or equilibrium stage was achieved at time intervals as shown in (Fig.5) against each discharge.

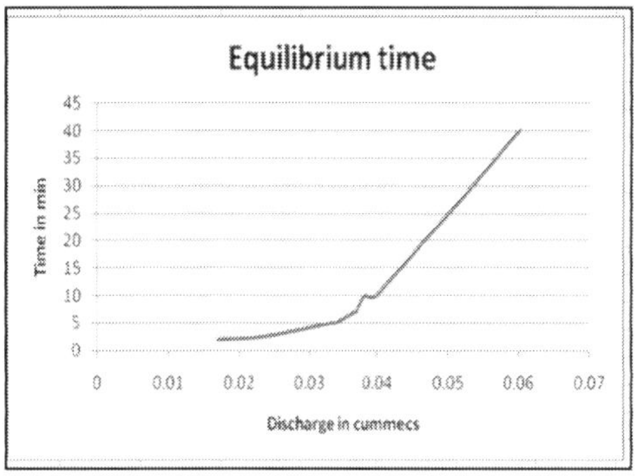

Figure 5 Equilibrium time against discharges

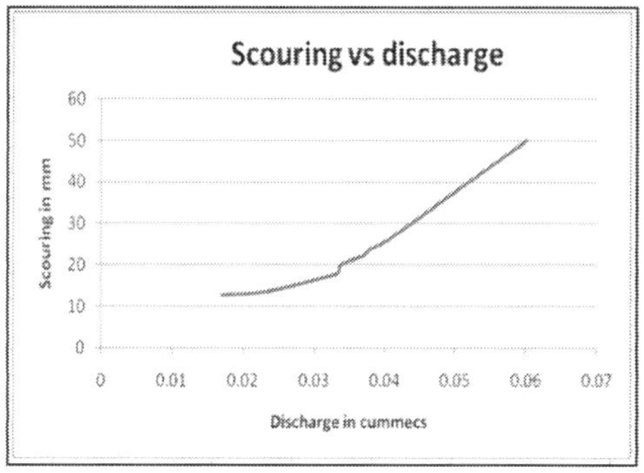

Figure 6 Scouring during 24 hours against discharges

- After conducting 10 experiments general scouring in bed was monitored and measured which was very minimal as shown in (Fig.6) against each discharge. However, after complete drainage of water, heavy scouring was observed at downstream of bridge model starting from 30.48 cm from piles and extending up to end of sand bed in the form of deep channel (Fig 7) each time pattern of channel formed was different.

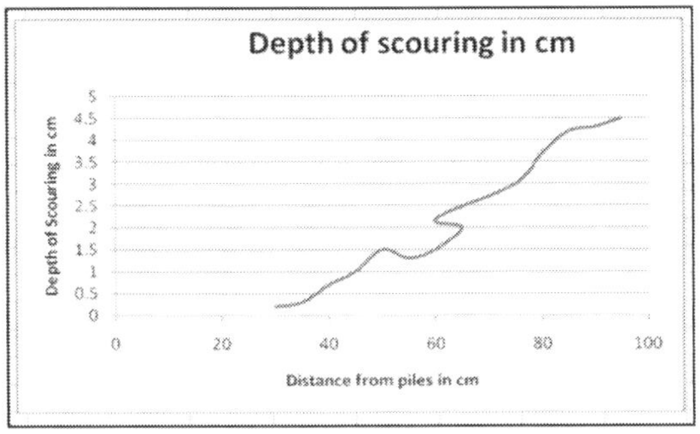

Figure 7 Scouring depth w.r.t distance from pile

Figure 8 Channels formation in bed (scoured river bed)

- The general scouring that occurred at downstream of model was nearly in same pattern as of original bridge (Fig 8). It was observed that scouring occurring at the time of draining off of water was more than live water scouring and same is the condition of harrow river flow at the downstream of Khanpur dam, Water flows through spillways of Khanpur dam only in rainy and high flood season and is main cause of general scour of channel bed.

- In order to reduce effects of scouring, scale down models of a baffle walls and weir were provided at different distances from piles and results were obtained. Baffle wall was not suitable due to its low height and heavy scouring occurs during experiments at upstream, downstream and sides of baffle walls.

- A weir at downstream show magnificent results and no further scouring was noted after piles were submerged initially there was live water scour but that was minimized with the lapse of time. After each experiment on weir water was allowed to drain off very carefully i.e. 3 days and very minute general and local scouring occurred that was 3mm to 4mm throughout the bed.

5. Conclusion:

Based on the laboratory experiments, this study after conducting 15 different experiments of different discharges concludes following

- Live bed scouring is higher for higher discharges.
- Scouring is higher at the time of draining off of water than live scouring of water.
- Downstream weir is beneficial for scour reduction even under high floods while baffle walls were of no importance.

6. Recommendations:

Provision of downstream weir narrows down scope of scouring in case of heavy flooding in inundation canal instead of baffle walls or any other measure as evident from failure of protection measures already provided in years 2000 and 2007. A storage weir construction at the site of bridge number 30 on Harrow River is recommended as best possible measure.

7. References:

[1] Mohammed, T. A., Noor, M. J. M. M., Ghazali, A. H., Yusuf, B., and Saed, K. (2007). Physical modeling of local scouring around bridge piers in erodable bed. *Journal of King Saud University-Engineering Sciences*, **19(2)**, 195-206.

[2] Farooq, R. and Ghumman, A. R. (2019). Impact assessment of pier shape and modifications on

scouring around bridge pier. Water, **11(9)**, 1761.
[3] Melville, B. W., and Coleman, S. E. (2000). Bridge scour. *Water Resources Publication.*
[4] Wang, Lu, et al. Effects of a downstream submerged weir on local scour at bridge piers. *Journal of Hydro-environment Research* **20** (2018): 101-109.

IOP Conf. Series: Materials Science and Engineering **899** (2020) 012016 doi:10.1088/1757-899X/899/1/012016

Sign Language Translation in Urdu/Hindi Through Microsoft Kinect

Sohail[1], M Y Khan[1], N A Siddiqui[2] and Kiran[3]
[1] Faculty of Computing, Mohammad Ali Jinnah University, Pakistan
[2] Department of Computer Sciences, Sukkur IBA University, Pakistan
[3] Department of Computer Sciences, Technical University Berlin, Germany

E-mail: sohailahmedhanfi@gmail.com

E-mail: yaseen.khan@jinnah.edu

Abstract. Communication is the foremost and basic necessity of a person. The ability to speak without any problem is a boon; nevertheless, woefully we have often seen people around us who are the victim of speech or hearing impairments. Thus, the 'Sign Language' (SL) appears as an alternate standard language that is globally understood and adopted by the deaf community. Though SL is a communication tool, but in practice, we still see most of the normal people do not understand the SL properly; thus, it is again a problem in communication between a speech/hearing impaired and normal persons. In this regard, many research attempts have been made to resolve this problem via wearable technology and other different paradigms of computational solutions. However, almost all of them focused on English or western languages, which provide a trivial resolution for the people of the Indian subcontinent. Thus, in this paper, we propose a solution to address this issue by employing Kinect Motion Sensor for detecting the signs (in SL) through hand movements and gestures and translating them in Urdu/Hindi language, which is readable and audible to normal persons. The proposed system is developed on a feature-based model; hence, we stored signs in a dictionary/training set for the development of the baseline system; however, the users can customize the system by adding their sign patterns in the future.

1. Introduction

Humans are exceptional amongst all living beings with the ability to speak, which permits vocalizing in order to communicate among themselves understandably. Thus a human who can speak, narrowly without flaw, can be characterized as a normal being. Regrettably, some people are not capable of speaking due to major deficiency, particularly with problems in the articulately or hearing organs, which makes them impaired to speak or hear [1] Hence, society has developed Sign Language (SL) as a medium of communication with subjects. The SL utilizes the gestures and movements of hands and fingers to precise the thought of subjects. The syntax and morphology of the SL play an important role, and the standard components (i.e., lip movements, facial expression, etc.) can produce a significant change in the meaning of a sign [2] However, these SLs differ from region-to-region as we see the change in natural languages worldwide [3]. The SL, although, offers a conversational medium among the subjects, but for the rest of the people, the situation is still the same.

In the recent era, we have seen computational treatments/solutions for overcoming the issue. The scientists and linguistics appraise various methods of Sign Language Translation (SLT) in this field. In the same regard, this paper focuses on the development of a tool-based approach for automated SLT. The underlying theme of the proposed work (as depicted in fig 1) is to employ the Microsoft Kinect sensor for identifying gestures in SL and then translating them into words that are supposed not only for displaying on-screen but also audible to normal people.

1.1. Research Goals and Objectives

The goal of this research is to develop an automated Sign language translator that uses the data provided by the Microsoft Kinect Xbox 360 TM sensor. Kinect works based on human joints as the X, Y and Z coordinates of these joints are moved Kinect catch them, and we can handle the results of motion in back end programming, An input sign given by the user is captured by Kinect camera and infrared sensor and translator provide the exact word in English and Urdu/Hindi language. One can easily comprehend the purpose of observing figure 1.

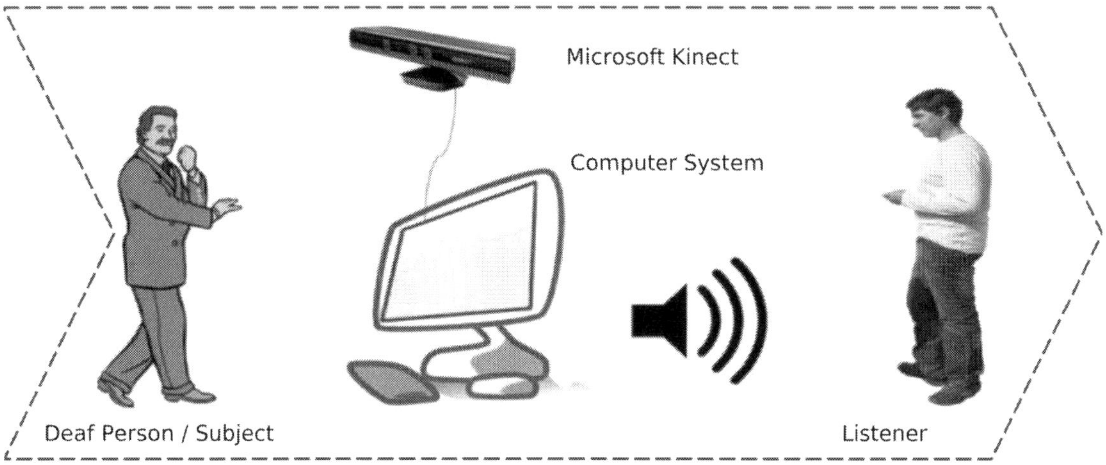

Figure 1: The theme of research work.

The work is done for this paper targets Indo-Pakistan SL as an input and Urdu/Hindi languages for the output language. Thus, with the two output languages (i.e., Urdu/Hindi which are mutually intelligible [4], and considered as the 3rd most spoken language w.r.t native speakers [5]), we address the problem for Indian subcontinent who have almost 7.4% of the disabled population in Pakistan [6] and 6.3% of the total population in India[7], respectively, are affected by hearing or auditory loss/impairment; which grosses around $\sim$ 99.8 million people (all together for both countries).

The other goals of this research work presented hereunto:

- Design a user-friendly interface so that users can easily understand and run the system.
- System should be capable of performing efficiently and produce results in run time so that the user can communicate in a real-time environment.
- Make user capable of storing its signs in sign dictionary for both languages, i.e., English and Urdu/Hindi. All of these features make a deaf person able to communicate with a normal person, which helps him to overcome his deficiency though it serves the society with positive outcomes.

2. Related Work

In Pakistan, Boltay Haath System [8] is a well-known gesture recognition system that is computer-based that converts signs produced by deaf into Urdu speech. Their technique works

IOP Conf. Series: Materials Science and Engineering **899** (2020) 012016 doi:10.1088/1757-899X/899/1/012016

by taking input as an incoming sign and compares it with a dataset of signs/gestures. They have got an accuracy level of nearly 70%.

Fuzzy classifier, which is another approach proposed by Kausar et al. [9], it used a classification algorithm based on fuzzy logic to recognize signs/gestures shown by the deaf. In their algorithm, an angle is measured for the classification of gestures between fingers and their joints. Accuracy of their work is 95%.

Haseeb et al. [10] built an inverse system named PSLIM for the deaf community in another way. In their system, the focus is on reducing the communication gap between the hearing impaired and hearing person. They convert sign/gesture of PSL by listening to the audio. They have achieved 78%.

Ali [11] came up with another solution that helped enhance communication between normal and deaf person . Their proposed system takes input as both text and image of a sign and translates it to another form. Their system is tested upon the Urdu alphabets of PSL using the Haar classifier with an RGB camera. However, not only the accuracy of the system is missing, but the nature of the dataset is too much complex for experimentation.

Khan et al. [12] successfully Converted PSL signs/gestures into Urdu language with the help of image processing . Their system provides easiness to allow deaf as a novice to learn computer programming and software engineering etc. It also proposed a comprehensive frame-work and used a subset of C++. The used RGB camera to capture 500 images of 37 letters of Urdu alphabets. Their technique works with skin region filtering using image processing. After that, they applied discrete wavelet transformation (DWT) by extracting features of interest. It is the first technique that is software-based without any hardware that is data glove etc. The presentation of their form is in textual form. They have selected 226 images out of 500 for training and 74% for testing and achieved accuracy 86%.

Sami et al. [13] recognized Urdu alphabets in Pakistan Sign Language (PSL). In their solution, computer vision technology was used for the recognition of PSL Urdu. They have found a close match between the input image and the image dataset. The accuracy of their work is achieved 75%.

Although a great deal of related work for SLT is mentioned above, the proposed work is uniquely based upon Microsoft Kinect 360 and allows the user to store his signs in sign dictionary to work for the Urdu and Hindi language.

3. Methodology

The proposed solution is divided into four main modules, as shown in the block diagram (see figure 2). The details of these modules are covered in the sub-sections 3.1 – 3.4, respectively.

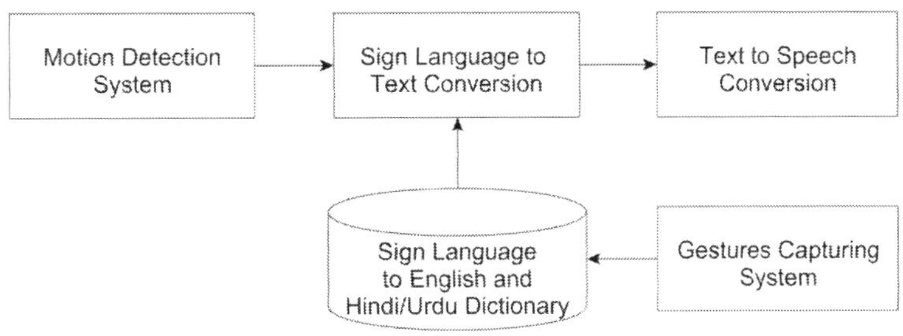

Figure 2: The architecture of the sign language translation system.

3.1. Human Gesture Detection
The Motion Detection System uses Microsoft Kinect for the Xbox 360TM device along with its SDK 1.7 to detect the signs of the deaf person; these signs consist of gestures that are performed by any deaf person in front of the Kinect. The Kinect consists of three basic parts that collaborate to sense any activity and produce the real image on the screen: an RGB color VGA video camera, a depth sensor, and a multi-array microphone [14]. The camera spots the facial expressions, body type, and color mechanism i.e., red, green, and blue. Its pixel resolution and frame rate are 640×480 and 30 fps, respectively. This aids in facial and body identification.

The depth sensor has a monochrome CMOS sensor and infrared projector that facilitates to form the 3D imagery all over the room. It also calculates the remoteness of each point of the user's body by transmitting invisible near-infrared light and measuring its 'time of flight' after it replicates the objects. These components can sense and trace 48 diverse points on each user's body and repeats 30 times every second. A famous algorithm of 'Dynamic time Wrapping (DTW)' has also been used to accurately identify the sign performed by an individual user while he is changing his position or multiple users are present in front of sensors.

3.2. Dynamic Time Wrapping (DTW)
This algorithm was presented in the 60s; it is modal to identify closeness between two instances [15], which may shift in time or speed. For example, the algorithm will recognize the pattern of a walking person regardless of whether in one instance the individual is walking in slow steps and if in another video he or she is walking fast, or regardless of whether there are increasing or decreasing velocities. By utilizing DTW, the system will have the capacity to locate an ideal match between two given instances (i.e., signs) with certain confinements. The instances are *wrapped* non-linearly in the time measurement to decide the measure of their likeliness In this research project, DTW has been utilized along with the input of Kinect sensor for motion/sign detection to identify the user even if he is changing the position more frequently and with different sign executions speeds.3.2. Sign language to English and Urdu/Hindi Dictionary The Sign language dictionary is a file containing all data about frames of gestures of joint X, Y coordinates corresponding to a sign, and the respective meaning in English and Urdu/Hindi language.

3.3. Sign to Text Conversion
The detected signs are compared with the signs already stored in the dictionary each time the Kinect sensor captures a frame. As the sign is marched, the sign to text converter results in the meaning of the sign in either English or Urdu/Hindi language, whatever the user selects language mode. A flat file has been used to store gesture coordinates and their corresponding meaning in English and Urdu language.

3.4. Text to Speech Converter
For English text-to-speech conversion, Microsoft's text to speech synthesizer library for .NET is used that takes English text as input and creates audio stream as output. It is also referred to as text-to-speech (TTS). A synthesizer must present an extensive analysis and processing to correctly alter a string of characters into an audio stream that seems just as the words would be uttered.

For Urdu/Hindi text-to-speech conversion, a database of own voices is created against every gesture in the dictionary, so when any gesture is detected, the system plays the voice of related word from the database in the Urdu language.

Users can capture and store his signs in the dictionary by using a simple gesture capturing system in the proposed solution. User is required to set himself in front of the Kinect sensor and see its skeleton; click on the capture button and write the name of sign that he wants to store the gesture takes 32 frames and store the sign performed in 3 seconds against the name of gesture given by the user.

3.5. Flow of Activities

The Activity starts with Deaf/Signer, who stays in front of the device, he will perform sign actions in the form of sign language Motions Detection system detects these sign by Using Kinect. The Sign to text system will check the sign language according to the Singer input; if signs are correct according to the dictionary of gestures, it will convert it into the English or Urdu text. The system will convert that text into English or Urdu audible form with the help of a speech synthesizer, and Listener will have the respectable voice of those signs which he/she performs.

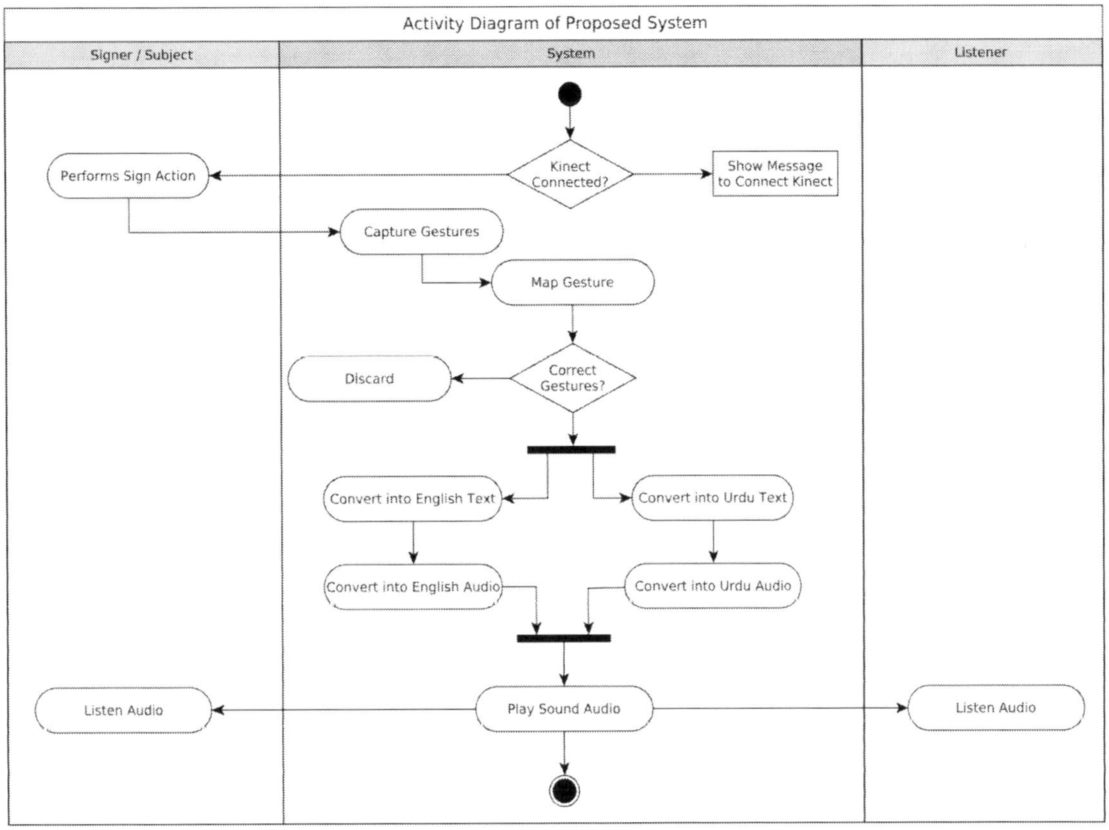

Figure 3: Activity diagram of the proposed system.

4. Result

A total of 24 signs have been stored in the sign dictionary, which are thoroughly tested with 10 subjects, alongside people without speech/hearing impairments. Since, it is a baseline idea, and a lot of sign words have to be employed, therefore, authors found the results with the currently stored signs are quite encouraging. The system successfully takes signs as input and convert it into English Urdu, and Hindi text and followed by conversion of the audio.

The authors also add that the paper has a limited space, therefore, a sample of most frequently experimented sign words are presented in table 1. Further, with the same reason of limit, we

IOP Conf. Series: Materials Science and Engineering **899** (2020) 012016 doi:10.1088/1757-899X/899/1/012016

only present the coordinate values for the word 'NO' in the table 2, to adjust the width of the table, the respective values are rounded off to 4 decimal places, however during the experiment the values were upto 15 decimal places.

Table 1: A sample of word list used in the proposed system.

English	Urdu	Hindi	Transliteration
I	مجھے	मुझे	/mʊdʒʰeː/
Help	مدد	मदद	/məd̪ əd̪/
Need	چاہیے	चाहिये	/tʃɑːhɪjeː/
Hello	ہیلو	हेल्लो	/heːloː/
Thanks	شکریہ	शुक्रिया	/ʃʊkrːrɪjɑː/
No	نہیں	नहीं	/nəhĩː/
Why	کیوں	क्यों	/kɪjũː/

Table 2: Coordinate values for word 'NO' in Indo-Pak SL, captured in Microsoft Kinect.

Microsoft Kinect Coordinate Values for 'No' Sign
⟨ -0.7769, -1.9738, -0.7909, -1.6615, -0.7178, -0.9049, 0.8706, -0.659, 0.7453, -0.1675, 0.7491, -0.0312 ⟩
⟨ -0.7768, -1.9742, -0.7904, -1.6612, -0.7178, -0.9047, 0.8707, -0.6605, 0.7525, -0.1512, 0.7487, -0.03 ⟩
⟨ -0.7758, -1.974, -0.7897, -1.6604, -0.717, -0.9048, 0.8628, -0.6561, 0.7494, -0.1721, 0.7505, -0.0422 ⟩
⟨ -0.7754, -1.9742, -0.7896, -1.6601, -0.7165, -0.9044, 0.8657, -0.6573, 0.7477, -0.1706, 0.7504, -0.0406 ⟩
⟨ -0.774, -1.9741, -0.7881, -1.6601, -0.7161, -0.9044, 0.867, -0.6585, 0.7478, -0.1704, 0.7509, -0.0372 ⟩
⟨ -0.7732, -1.9741, -0.7874, -1.6605, -0.7153, -0.9048, 0.8641, -0.6583, 0.7508, -0.1735, 0.7513, -0.037 ⟩
⟨ -0.7732, -1.9744, -0.7874, -1.6604, -0.7149, -0.9045, 0.8659, -0.6605, 0.7502, -0.1733, 0.7518, -0.0392 ⟩
⟨ -0.7731, -1.9746, -0.7873, -1.6611, -0.7148, -0.9051, 0.8667, -0.6617, 0.7505, -0.1735, 0.7535, -0.0439 ⟩
⟨ -0.773, -1.9756, -0.7874, -1.6615, -0.7147, -0.9052, 0.8683, -0.6626, 0.7494, -0.1725, 0.7535, -0.0414 ⟩
⟨ -0.7724, -1.9745, -0.7867, -1.6608, -0.7143, -0.905, 0.8678, -0.663, 0.7485, -0.1732, 0.7525, -0.0469 ⟩
⟨ -0.7724, -1.9749, -0.7866, -1.6609, -0.714, -0.9046, 0.8687, -0.6643, 0.7496, -0.171, 0.7531, -0.0488 ⟩
⟨ -0.7727, -1.9753, -0.7864, -1.6612, -0.7139, -0.9049, 0.8688, -0.6649, 0.75, -0.1712, 0.7528, -0.0472 ⟩
⟨ -0.7724, -1.974, -0.7864, -1.6601, -0.7137, -0.9044, 0.8679, -0.6652, 0.7493, -0.171, 0.7511, -0.0441 ⟩
⟨ -0.7719, -1.9745, -0.7867, -1.6602, -0.7137, -0.9046, 0.8684, -0.6662, 0.7494, -0.1715, 0.7514, -0.0451 ⟩
⟨ -0.7725, -1.9757, -0.7872, -1.6609, -0.714, -0.9046, 0.8686, -0.6668, 0.7507, -0.1707, 0.7522, -0.0424 ⟩
⟨ -0.773, -1.9774, -0.7875, -1.6628, -0.7141, -0.9061, 0.8702, -0.6688, 0.7511, -0.1719, 0.7541, -0.0499 ⟩
⟨ -0.773, -1.9781, -0.788, -1.6623, -0.7137, -0.9062, 0.87, -0.6692, 0.7508, -0.1724, 0.7543, -0.053 ⟩

5. Conclusion

We conclude that our proposed system, i.e. for the translation of Indo-Pak sign language in respective speech/audios, shows positive results w.r.t available signs in user-defined dictionary. The system is also capable to grow the dictionary through the addition of signs. Although the data is limited for this research work, but it proves the potential of the proposed methodology. This research work provides an opportunity for the development of sign language translator in different languages. Alognside expanding the sign-dictionary, we aim to provide this solution for portable devices in the future so that deaf people can carry the system with them and get more benefit from the system.

IOP Conf. Series: Materials Science and Engineering **899** (2020) 012016 doi:10.1088/1757-899X/899/1/012016

References

[1] Deafness and hearing loss - world health organization https://www.who.int/news-room/fact-sheets/detail/deafness-and-hearing-loss accessed: 2020-04-10
[2] Elliott E A and Jacobs A M 2013 *Frontiers in psychology* **4** 115
[3] Hassan B, Farooq M S, Abid A and Sabir N 2015 *VFAST Transactions on Software Engineering* **9** 1–6
[4] Khan M Y and Nizami M S 2020 *2020 International Conference on Information Science and Communication Technology (ICISCT)* (IEEE)
[5] Khan M Y, Emaduddin S M and Junejo K N 2017 *2017 IEEE 11th International Conference on Semantic Computing (ICSC)* pp 242–249
[6] Niazi F With sign language, i am equal - the news https://www.thenews.com.pk/amp/152779-With-Sign-Language-I-am-Equal accessed: 2020-04-10
[7] Varshney S *et al.* 2016 *Indian journal of otology* **22** 73
[8] Alvi A, Azhar M, Usman M, Mumtaz S, Rafiq S, Rehman R and Ahmed I 2007 *International Journal of Computer, Electrical, Automation, Control and Information Engineering, World Academy Of Science, Engineering and Technology* **1**
[9] Kausar S, Javed M Y and Sohail S 2008 *Proceedings of the 8th conference on Signal processing, computational geometry and artificial vision* (World Scientific and Engineering Academy and Society (WSEAS)) pp 101–105
[10] Haseeb A A, Illyas A J and Ilyas A 2012 Speech translation into pakistan sign language
[11] Ali S A Detection of urdu sign language using harr algorithms
[12] Khan N S, Abid A, Abid K, Farooq U, Farooq M S and Jameel H 2015 Speak pakistan: Challenges in developing pakistan sign language using information technology
[13] Sami M, Ahmed H, Wahid A, Siraj U, Ahmed F, Shahid S and Shah S I A 2014 *2014 International Conference on Robotics and Emerging Allied Technologies in Engineering (iCREATE)* (IEEE) pp 200–204
[14] Zhang Z 2012 *IEEE MultiMedia* **19** 4–10
[15] 2007 *Dynamic Time Warping* (Berlin, Heidelberg: Springer Berlin Heidelberg) pp 69–84 ISBN 978-3-540-74048-3 URL https://doi.org/10.1007/978-3-540-74048-3$_4$

IOP Conf. Series: Materials Science and Engineering **899** (2020) 012017 doi:10.1088/1757-899X/899/1/012017

Developing Non- Linear Relationship Among Factors Affecting the Rutting Susceptibility of Asphalt Mixtures Using Two Parameter Weibull Distribution

Fatima Ashfaq[1], Imran Hafeez[2] and Sabahat Hussan[3]
Department of Civil Engineering, University of Engineering & Technology, Taxila, Pakistan

1) fatima.uettaxila@gmail.com, 2) imranhafeez783@yahoo.com

Abstract. This paper illustrates a novel approach to determine the probabilistic rutting in asphalt mixtures based on Weibull distribution statistical approach. In this paper, thirty wearing course asphalt mixtures prepared using different binder grades and aggregate sources, are analysed. Cooper Wheel tracker Test (CWTT)and Asphalt pavement analyser (APA) have been used to ascertain the rutting potential of asphalt mixtures with varying binder grade and aggregate source. Mixtures are then ranked based on their rutting susceptibility. The obtained data is processed using two parametric Weibull approach. Scale and shape parameters are figured out by multiple regression technique. Test results are analysed using goodness of fit test. Results indicate that rut depth predicted from two parameter Weibull distribution technique corelates well with the actual rut depth with R^2 value of 0.88 for both cooper wheel tracker and asphalt pavement analyser test. It has been found that Weibull distribution bests fit the NHA-A, NHA-B and SP-B graded mixtures with NRL 60/70 at 50◦C. A good correlation has been observed between Cooper Wheel tracker and Asphalt pavement analyser as translated by R2 of 0.86.This research is important in a sense that several agencies have adopted rutting as a failure criteria for flexible pavements. By mathematically modelling the rutting response, predicting the rutting behaviour of mastics becomes simpler and more efficient in approach.

1. Introduction:

Rutting is defined as an accumulation of unrecoverable strain resulting from applied load which is visible as depressed channel in wheel paths of asphalt pavement. Major and minor changes accumulated during the design life of pavement that eventually depreciates the pavement performance[1]. There are number of factors that affect the functional as well as structural rutting performance of asphalt mixtures. Deceleration and repeated loads from heavy traffic can create pronounce ruts in wheel path of asphalt concrete pavement [2]. Asphalt mixture properties i.e. binder type and aggregate gradations in addition to compaction efforts significantly affect the rutting potential[3].Sabahat et. al,[4] investigated the factors effecting permanent deformation of asphalt mixtures using Asphalt pavement analyzer (APA) and concluded that temperature variable most significantly affects the rut depth[5]. Timely prediction of pavement response is critical to improve the design of new and existing pavements. Moreover, laboratory rutting tests show scatter results even by adopting high quality control and testing practices[6]. Therefore, it is required to move towards mathematical based pavement performance models that will reduce the testing effort and predict permanent deformation with more accuracy. As this distress has been associated with majority of flexible pavements, so there is a need to develop simple mechanistic empirical techniques in-parallel to ascertain the rut potential of asphalt mixtures and to reduce the time and efforts required in performing such tests in laboratory. A handful of studies are available that present different techniques of simulating laboratory rutting to actual in-field rutting. The models were developed to timely predict the pavement performance.

Haydari[7] modeled the rutting of asphalt pavements obtained from uniaxial creep test using Artificial Neural Network (ANN) and Adaptive Neuro-Fuzzy (ANFIS) techniques. XU and

Muhammad[8] developed a Mechanistic Empirical (ME) approach to model asphalt pavement rutting where the rutting is empirically correlated to the cyclic deflection which is evaluated using finite element modelling. Monismith[9] used a regression-based approach to describe the rutting behavior of asphalt mixtures on Wes track test sections. Gao et; al,[10] employed the solution of Burger's model to predict rutting and compared the final model with Finite Element Method (FEM), which validated the model pretty well. Collop et al,[11] studied the rutting behavior based on its viscoelastic behavior using VESYS software. Wong and Suo[12] established a nonlinear regression model by performing dynamic cyclic compression test. The development process and models of repeated load axial permanent deformation test was presented by Boateng and Maina[13] in the form of non-linear regression equation using a standard South African hot-mix asphalt mixture. Zhou et al,[14] proposed a three-stage model to accurately characterize three stages of permanent deformation. Gandomi,[15] concluded that flow number is an explanatory index to predict rutting potential of asphalt mixtures by utilizing a gene expression programming (GEP).
Of many techniques adopted, Weibull distribution is a widely used statistical technique in reliability and life data analysis due to its versatility and simplicity. The family of Weibull distribution was introduced in 1939 by a Swedish physicist Waloddi Weibull. It is one of the best known life time distributions which can adequately model various types of observed failures[16]. Two parameter Weibull Distribution requires shape parameter (β) and scale parameter (η) for reliability analysis. In order to evaluate whether data could be modelled using Weibull distribution, the best way is to use statistical software or by manually plotting data on Weibull probability plot. A straight line on Weibull probability plot is an indication that data follows Weibull distribution. There are limited number of studies conducted to date to model the laboratory rutting performance using Weibull Distribution technique as shown in Table 1.

Table 1: Literature review on modelling laboratory rutting using Weibull distribution technique.

References	Year	Materials Used	Experimental Technique	Results
Bala and Napiah[17]	2018	Modified asphalt mixture with polypropylene and polyethylene in addition to non-silica particles	Wessex Wheel Tracking Device	A good correlation was found between measured and laboratory rut depth.
Singh and Krishna[6]	2018	Asphalt mixture	Flat Rubber Wheeled Loaded Wheel Tester	Weibull distribution resulted in conservative rut results while lognormal presented overestimated rut curves
A.S.M Asif ur Rehman, Matias M. Mendez Larrain and Rafiqul A. Tarefde[18]	2017	Hot Mix Asphalt (HMA)	Hamburg Wheel tracking Device	The developed regression based predictive model using Weibull Distribution is found to be a good predictor of in-situ rutting.
Matias M. Mendez Larrain and Rafiqul A. Tarefde[19]	2016	Warm Mix Asphalt (WMA)	Hamburg Wheel Tracking Device	Weibull Function is a good estimator of lower rutting depths and where stripping phase has not been analyzed.

Coleri et al [20]	2008	Asphalt Concrete (AC) mixtures	Repeated simple shear test at repeated height (RSST-CH).	Integrated Weibull approach is a good simulator of stage I and II of in-situ rutting of asphalt mixtures.

In past literature, asphalt mixtures were prepared using single batch of binder grade and aggregate samples. These mixtures were then used to test the rutting performance of asphalt mixtures. Results obtained from single wheel tracking device were then simulated using Weibull distribution technique. In this study, extensive data has been collected to model the rut behavior. Three binder grades and aggregate samples were used to prepare asphalt mixture using Marshal Mix design. Performance testing was then conducted on Cooper Wheel Tracking Device and Asphalt Pavement Analyzer (APA). This wide results in terms of mixture variability and testing techniques were then simulated with modelled Weibull Distribution technique. One more modification in this research study is the correlation of rut results obtained from CWT and APA using non-linear regression equation. The main objective of this research study is to develop a probabilistic approach to predict the true rutting curve that will provide enough flexibility to finalize the pavement design based on type of materials being used and testing conditions being adopted at the field.
The contributions of this research study are threefold:

1. To investigate the rutting behaviour of asphalt mixtures composed of variety of bituminous and aggregates mixtures.
2. To develop relationship among various parameters effecting the rutting of asphalt mixtures based on CWT and APA using two parameter Weibull Distribution.
3. Comparing CWT and APA test techniques with two parametric Weibull distribution to verify the simulation results.

2. Methodological Approach:

A total of thirty wearing coarse mixtures with varying aggregate and bitumen sources have been considered for this research study. Each mixture is unique in its own characteristics based on type of bitumen and aggregate used. Figure 1 gives a summary of methodological approach adopted for modelling the rutting behavior of asphalt concrete mixtures.

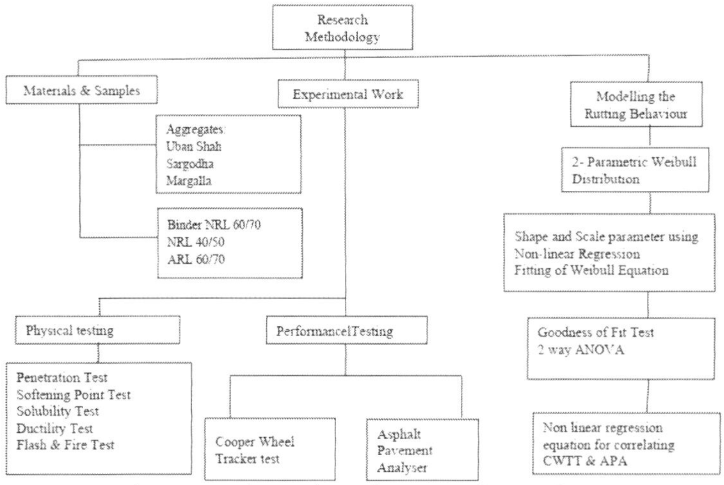

Figure 1: Flow Chart for Experimental Program

For each asphalt mixture, properties of individual materials were evaluated first and then a full material characterization was performed.

3. Materials Used:

3.1 Aggregates:

The properties of aggregates play an important role in overall performance of asphalt mixtures as 90% of asphalt concrete mixture consists of aggregates. For this research study, aggregates selected from Uban shah quarry of Sindh province, Sargodha quarry and Margalla quarry located in Punjab Province were analyzed. Flakiness and Elongation index test, sand equivalent test, water absorption test, soundness test and los Angeles abrasion value tests were performed to assess the performance of aggregates in HMA mixture. Aggregates queried from Uban shah were found to have flakiness index of 12% and Elongation index of 17% which is more than the specified limit of 10% according to BS 812.108. While Sargodha and Margalla aggregates had less proportions of flaky and elongated particles. Uban Shah, Margalla and Sargodha aggregates were qualitatively tested by performing sand equivalent test (ASTM D2419), Los Angeles abrasion value test (ASTM C131), Water absorption (ASTM C127), soundness test (ASMT C88) and Uncompacted void test (ASTM C1252). It has been found that all aggregate samples fulfil the standard specified limit prescribed in their respective ASTM standards. These tests translated the quality performance of different aggregate samples used and NHA-A gradation aggregates were found to be more tough and stronger as compared to other ones. While aggregates queried from Uban Shah were found to be more prone to loading and other physical and environmental condition due to presence of large number of flat and elongated particles.

3.2 Bitumen:

The bitumen sources investigated in this research study i.e. NRL 60/70, NRL 40/50, ARL 60/70 were manufactured from National Refinery Limited, Karachi and Attock Refinery Limited. Softening point test (ASTM D36), Penetration test (ASTM D5), Ductility test (ASTMC88) and Flash and fire point test (ASTM C142) were carried out to determine the viscosity, penetration index and safe temperature range for mixing asphalt. NRL 40/50 was found to have softening point of 49°C, followed by ARL 60/70 having 48°C and NRL 60/70 with 46°C softening point. Penetration value of NRL 60/70 is found to be 70mm, followed by ARL 60/70 with 63mm and NRL 40/50 with 49mm. Highest penetration values along with lowest softening point value of NRL 60/70 indicates increased viscosity as compared to NRL 40/50 and ARL 60/70.

4. Asphalt Mixture Design:

Marshall Mix design has been adopted for preparation of asphalt concrete specimens according to standard ASTM D 1559 specifications. A total of fifteen samples including three trials of asphalt-aggregate binder blends with five samples for each blend were prepared for each mixture. Results of Marshall mix design are tabulated in Table 2.

Table 2: Marshall Mix Design

Aggregates	Gradation	OBC	Stability
Uban Shah NRL 60/70	NHA-A	3.93	1190
	NHA-B	4.65	1198
	SP-A	5.40	1160
	SP-B	4.55	1182
	MS-II	4.70	1375
Uban Shah NRL 40/50	NHA-A	3.78	1189
	NHA-B	4.60	1006
	SP-A	5.50	1176
	SP-B	4.44	1073
	MS-II	4.81	1255

Margalla AL 60/70	NHA-A	4.00	1280
	NHA-B	4.10	1315
	SP-A	5.00	1270
	SP-B	4.40	1295
	MS-II	4.80	1340
Margalla NRL 40/50	NHA-A	3.88	1298
	NHA-B	4.43	1250
	SP-A	5.60	1383
	SP-B	4.50	1214
	MS-II	4.73	1586
Sargodha ARL 60/70	NHA-A	4.00	1260
	NHA-B	4.60	1225
	SP-A	5.60	1015
	SP-B	4.50	1180
	MS-II	5.00	1064
Sargodha NRL 40/50	NHA-A	3.90	1266
	NHA-B	4.50	1255
	SP-A	5.50	1028
	SP-B	4.40	1176
	MS-II	4.90	1086

Marshall Mixture design summary has been tabulated in Table 1 which indicates the results for 30 wearing coarse mixtures using NRL 40/50, NRL 60/70 and ARL 60/70 bitumen samples and Uban Shah, Margalla and Sargodha aggregates.

It can be inferred from Table 1 that asphalt mixtures prepared from Uban Shah aggregates have low stability values as compared to asphalt mixtures prepared from other aggregates. As Margalla aggregates and NRL 40/50 binder, when qualitatively tested, performed better as compared to other samples. This is the reason that asphalt mixtures prepared with Margalla aggregates and NHA 40/50 binder showed highest stability values as referred to Table 1.

Preparation of test specimens:

Research study has been conducted in two phases. In phase I of study, Asphalt pavement analyzer has been used to test the rutting of asphalt mixtures that are prepared from super pave gyratory compactor. APA is a laboratory method for an accelerated evaluation of permanent deformation. Rutting of mixes is assessed by placing a cylindrical sample under repetitive loading of 534N. APA features an automated data acquisition system that displays rut measurements in graphical format According to the standard protocol. Based on past literature, 5mm rut depth or 8000 loading cycles, whichever comes first, is considered as pass-fail criteria. A study was conducted at NCAT by Kandhal et al,[21] and in Florida by Choubane et al.[22] to evaluate the rutting potential of asphalt mixes using APA. Results indicated that APA has a fair potential of simulating the in-situ rutting of pavement.

Phase II of research considered the use of Cooper Wheel tracker (CWT) device to predict permanent deformation of samples compacted on roller compactor. Rutting measured by rolling a small loaded wheel device of 720N repeated over the specimen is then co-related to actual in-service pavement rutting.

5. Weibull Distribution:

The Weibull distribution in case of fitting CWTT and APA rut depth presented as the function of number of load cycles can be represented by hazard/failure rate function. The failure rate function of Weibull distribution function is expressed as follows;

$$R(N) = \frac{\beta}{\eta}\left(\frac{N-\gamma}{\eta}\right)^{\beta-1} \text{ Equation} - 1 \qquad (1)$$

Where R(N) is the deformation rate in "mm" at 'N' number of cycles and loading conditions, β is the shape factor also known as slope parameter and η is the scale factor. For two-parameter Weibull distribution, location parameter "y" is not used, and this value can be set to zero. Other than shape and scale parameters, Rut depth is also dependent on the number of loading cycles which is one the most important factor in modelling rut prediction models[23]. Successful fitting of Weibull distribution to rut data would mean that thorough understanding of rutting behaviour can be ascertained without doing prior laboratory destructive testing.
Shape parameter shows the effect of how deformation rate increases or decreases in a function and represented by the shape parameter bath curve as shown in Figure 2

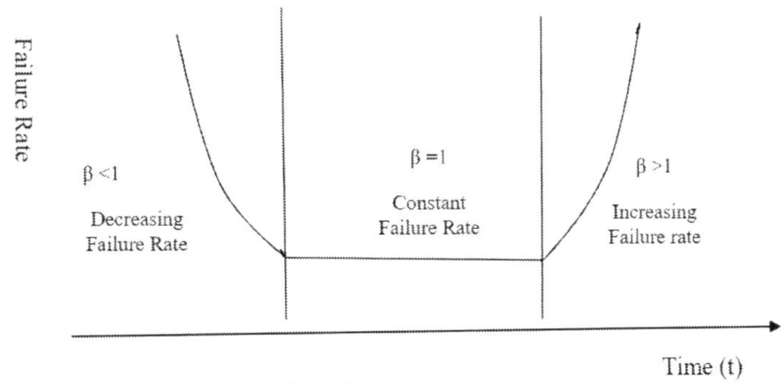

Figure 2:Weibull Shape parameter Bathcurve

Weibull function with β <1 has a distorting rate that decreases with time, also known as early failure life whereas function with β >1 has a deformation rate that increases with time. Weibull function with β =1 translates constant rate of deformation. A change in the scale parameter "η", has an effect of stretching out the function.

6. CWTT and APA rutting test results:

Total thirty wearing coarse mixtures with varying bitumen and aggregate sources were prepared using roller compactor and tested on APA and Cooper Wheel tracker to assess the rutting propensity of these asphalt mixtures. Figure 3 shows output results of CWTT test while Figure 4 shows the APA rutting results at temperature conditions of 40°C and 50°C.

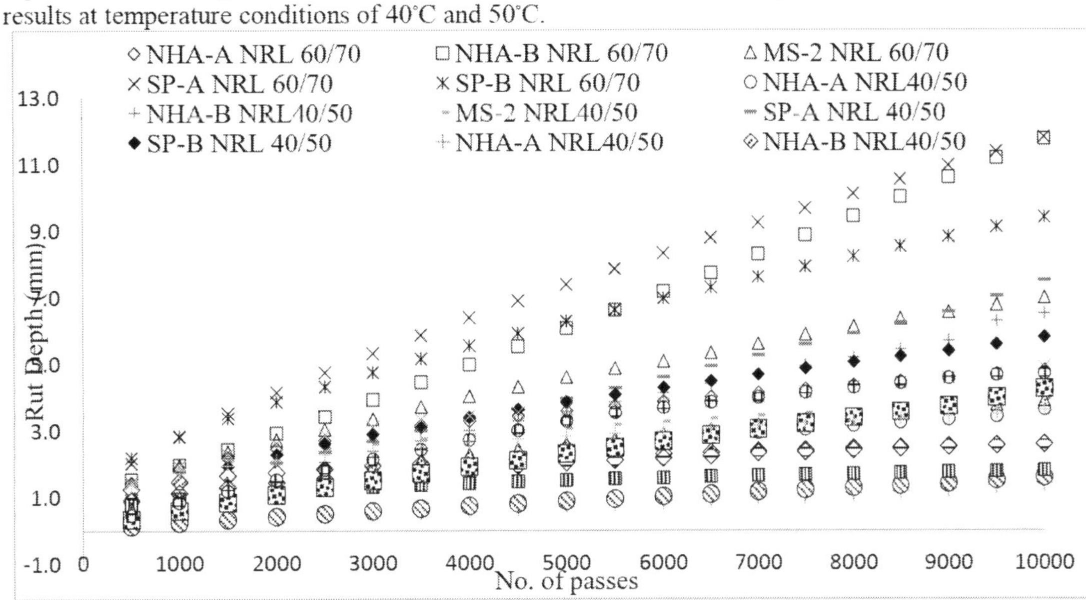

(a)

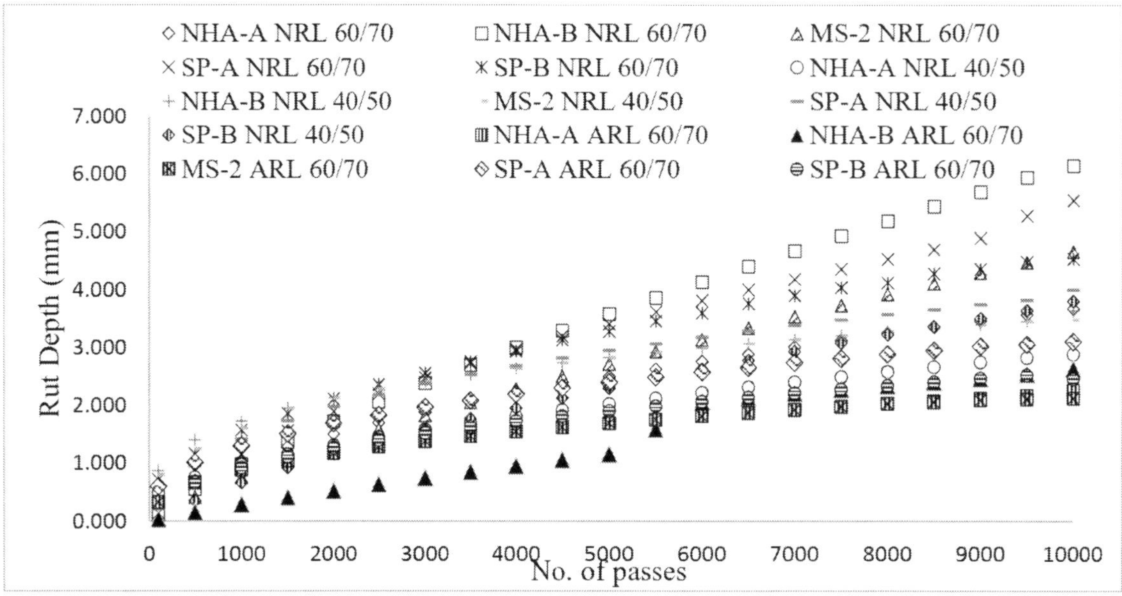

(b)

Figure 3: CWTT rut curves for different HMA mixtures at (a) 50°C (b) 40°C

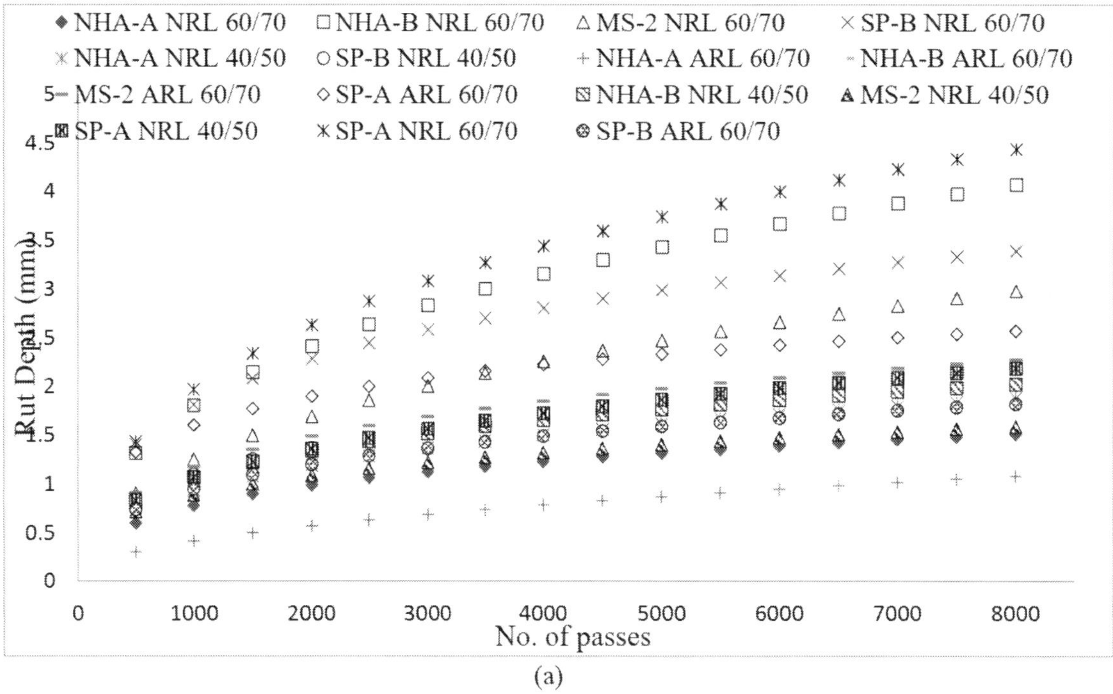

(a)

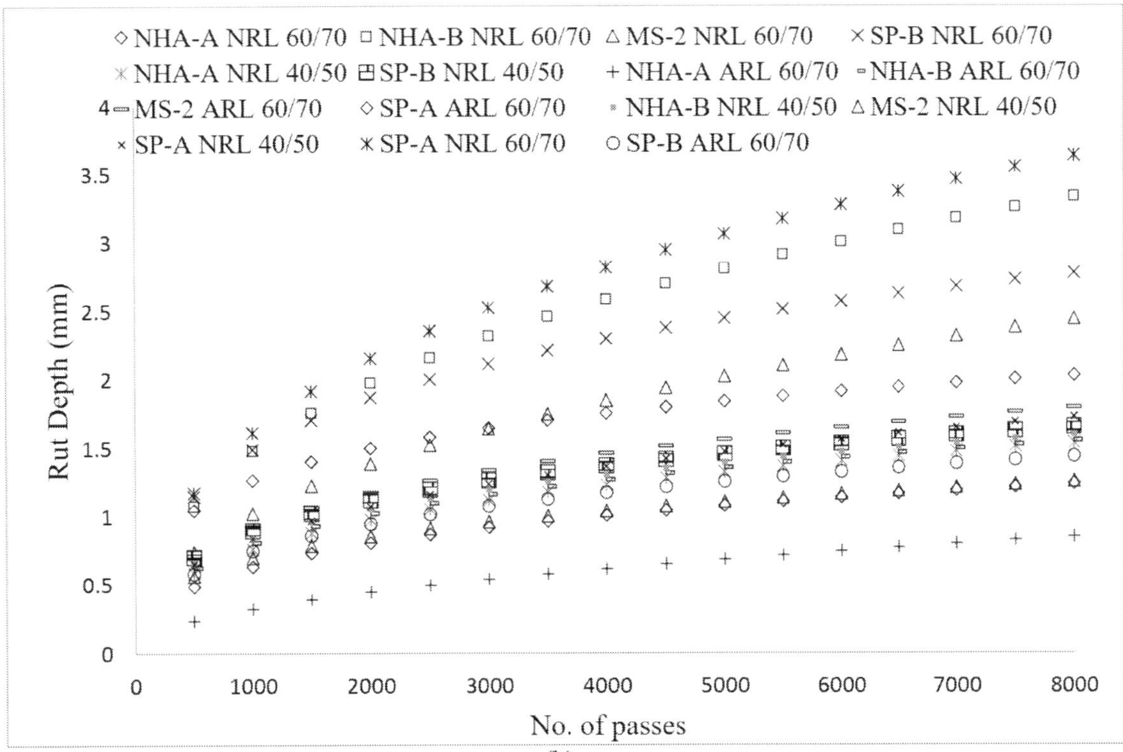

(b)

Figure 4: APA rut curves for different HMA mixtures at (a) 50°C (b) 40 °C

Test results indicate that mixtures are more susceptible to rutting at temperature condition of 50°C as compared to 40°C for both CWTT and APA mixtures as shown in Figure 3 (a),(b) and Figure 4 (a),(b). It is also been observed that SP-A NRL 60/70 mixtures are more susceptible to rutting as compared to other mixtures because of the presence of flaky and elongated aggregate particles. Also, NRL 60/70 has high penetration index as compared to bitumen grade 40/50. No sign of stripping has been observed for AC mixtures up to the limiting value of 10,000 loading cycles for CWTT and 8000 cycles for APA mixtures.

7. Weibull Probability Plots:

Weibull probability plot is a graphical technique employed to ascertain whether data set logically fits 2-parameter Weibull distribution at first place. Probability plots show a graph with observed cumulative percentage on abscissa and expected cumulative percentage on ordinate. Scatter plots close to reference line translates that data set follows the Weibull distribution. There are number of ways to plot Weibull probability plot. For this research study, Weibull-DR reliability software has been used to plot probability graphs for all asphalt mixtures. Figure 5 (a), (b) shows the sample Weibull probability plots for CWTT and APA.

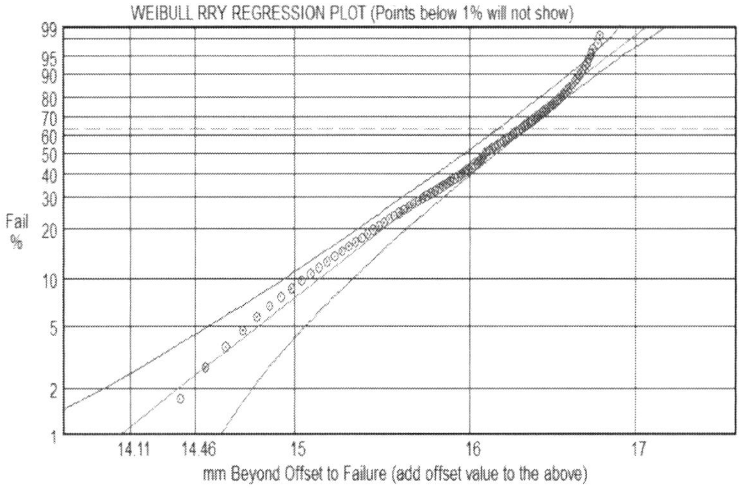

(a)

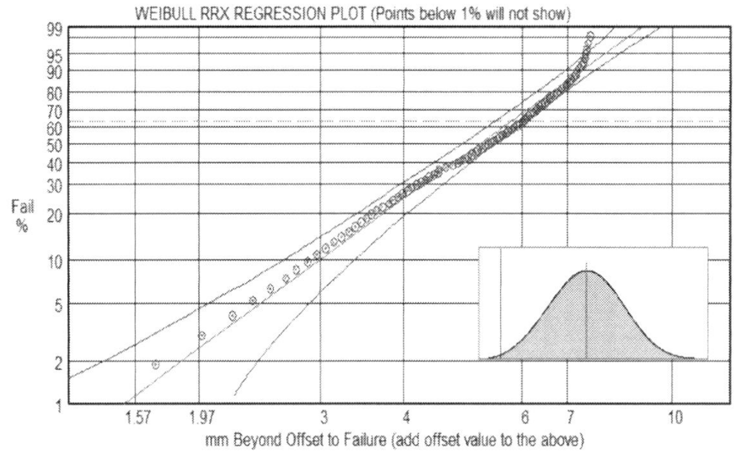

(b)

Figure 5: Weibull Probability Plot for (a) CWTT and (b) APA

Weibull plots has special scale designed so that if data follows Weibull distribution, points will be linear as can be shown on Figure 5. The least square fits of this line yield the scale and shape parameter of Weibull distribution. Mathematically, the shape and scale parameters can be figured out from probability plot as shape parameter is the reciprocal of slope of fitted line and scale parameter is exponent of intercept of fitted line. Table 3 shows the estimated values of shape and scale parameters for all asphalt mixtures.

Table 3: Estimated Values of shape and scale parameters

Asphalt Mixture	Gradation	β	η	Se/Sy	R^2	RMSE	n
Uban Shah NRL 60/70	NHA-A	1.34	4	0.1	0.998	2.4	100
	NHA-B	1.51	5.80	0.08	0.992	1.7	100
	SP-A	1.57	8.00	0.1	0.998	0.18	100
	SP-B	1.51	6.74	0.1	0.997	1.24	100
	MS-II	1.36	3.40	0.1	0.993	5.43	100
Uban Shah NRL 40/50	NHA-A	1.29	4	0.1	0.984	0.63	100
	NHA-B	1.3	3.9	0.1	0.996	0.74	100
	SP-A	1.35	4	0.1	0.964	1.50	100
	SP-B	1.275	3.3	0.1	0.997	0.67	100
	MS-II	1.31	4.1	0.1	0.993	0.16	100
Margalla ARL 60/70	NHA-A	1.29	3.40	0.1	0.995	0.44	100
	NHA-B	1.31	3.3	0.1	0.999	2.02	100
	SP-A	1.3	3.22	0.1	0.987	0.29	100
	SP-B	1.28	3.30	0.1	0.999	2.07	100
	MS-II	1.28	3	0.1	0.997	0.19	100
Margalla NRL 40/50	NHA-A	1.29	3.6	0.1	0.947	0.45	100
	NHA-B	1.38	3.90	0.1	0.998	2.12	100
	SP-A	1.37	3.50	0.1	0.998	0.10	100
	SP-B	1.36	4.02	0.1	0.999	0.01	100
	MS-II	1.28	3.18	0.1	0.993	0.17	100
Sargodha ARL 60/70	NHA-A	1.20	4.00	0.1	0.994	1.69	100
	NHA-B	1.25	3.7	0.1	0.995	0.24	100
	SP-A	1.28	4	0.1	0.988	1.17	100
	SP-B	1.21	3.50	0.1	0.996	0.13	100
	MS-II	1.24	4	0.1	0.968	1.09	100
Sargodha NRL 40/50	NHA-A	1.28	3.60	0.1	0.997	1.10	100
	NHA-B	1.29	3.40	0.1	0.979	2.84	100
	SP-A	1.30	3.50	0.1	0.973	1.98	100
	SP-B	1.28	3.40	0.1	0.996	1.09	100
	MS-II	1.30	3.90	0.1	0.989	0.43	100

8. CWTT and APA rut prediction model:

The Weibull distribution is one of the most commonly used distributions in reliability analysis because of its ability to take on various forms by adjusting its parameters. Previous researches have resulted that temperature has a very pronounced effect in rutting of asphalt mixes[3]. The relationship between shape factor "β", temperature "T" and asphalt mixtures' Stability "S" was determined by performing multiple regression with shape parameter being considered as dependent variables and temperature and stability is considered as independent variable.

Equation 2 shows the results of fitting a multiple linear regression model to describe the relationship between dependent variable and 2 independent variables. The equation of the fitted model is;

$$\beta = 2.0759 - 0.3123 \times \log(S) + 0.0034 \times T \quad \ldots\ldots\ldots\ldots\ldots\ldots\ldots \quad 2$$

As above equation shows that β is the function of temperature and stability of asphalt mixtures. Figure 6 (a) and (b) shows the effect of stability and temperature variations on the β for CWTT and APA respectively. Results have been graphically presented for both CWTT and APA. Stability has been plotted as an abscissa while β is plotted as an ordinate.

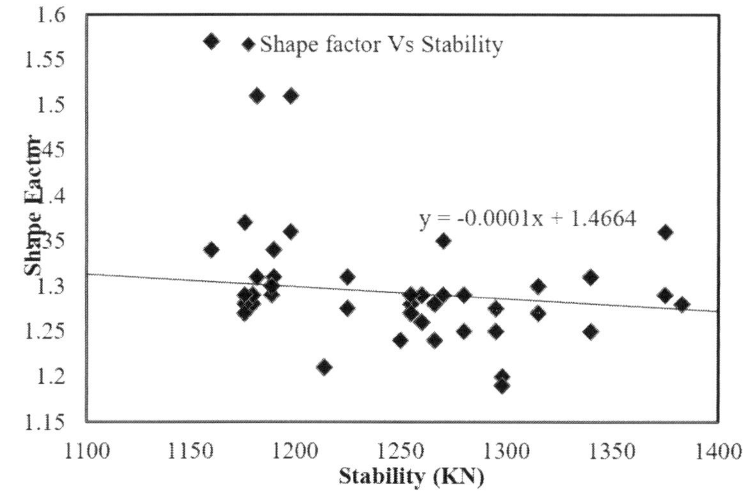

(a)

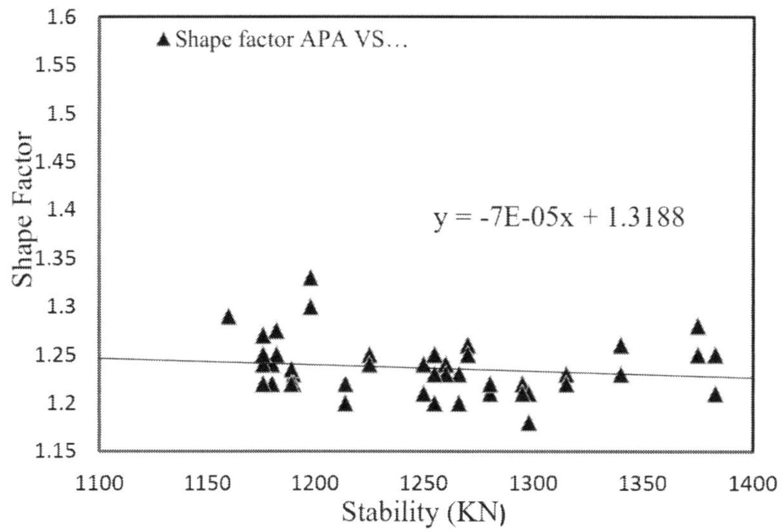

(b)

Figure 6: Effect of Stability on Shape parameter (a)for CWTT (b) for APA

Stability is an indicator of collective properties of aggregates and bitumen in an asphalt mixture. As it can be seen from Figure 5 that β found for the mixtures are greater than 1.0. This indicates that mixtures are experiencing normal/increasing wear out failure rate with time. Figure 6 shows the effect of temperature variation on β for CWTT and APA data. The trend shows that temperature has a direct relation with β. As temperature increase from 40 to 50°C, value of β gradually increases.

Figure 7 shows the effect of temperature variation on the shape parameter for CWTT and APA data. The trend shows that temperature has a direct relation with the shape parameter. As temperature increase from 40 to 50°C, value of shape parameter gradually increases.

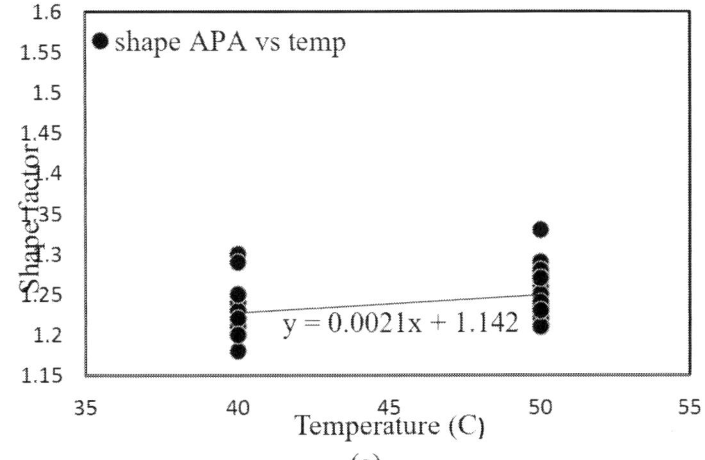

(a)

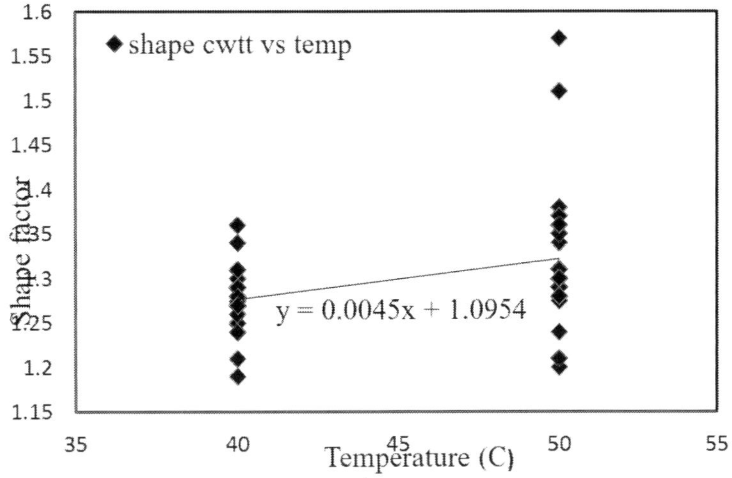

(b)

Figure 7: Effect of Temperature variation on Shape parameter (a)for APA (b)for CWTT

This correlation depicts that increase in temperature will increase the value of β which would then affect the rut depth, as equation 1 shows a direct relation between rut depth and β. Moreover, stability has an indirect relationship with β. So, increase in stability will in turn slows down the deformation rate in asphalt mixtures.

Based on the statistical approaches adopted in the past literature, it is considered that scale factor "η" is always dependent on the shape (β) and location factor (y). As location factor is taken as zero, η is entirely dependent on β. Equation 3 shows the relationship obtained as the result of multiple regression technique. The output shows the results of fitting a multiple regression model to describe the relationship between scale (η) and shape (β) factor where η is taken as dependent variable and β as an independent variable. The equation of the fitted model is

$$\eta = -1.1 + 3.9877 \times \beta \quad \ldots\ldots\ldots\ldots\ldots\ldots \quad 3$$

Equation 3 is graphically presented in Figure 7 to exhibit the relationship between dependent and independent variables. Test results of CWTT and APA data both has been plotted in the Figure 7 (a) and (b), respectively.

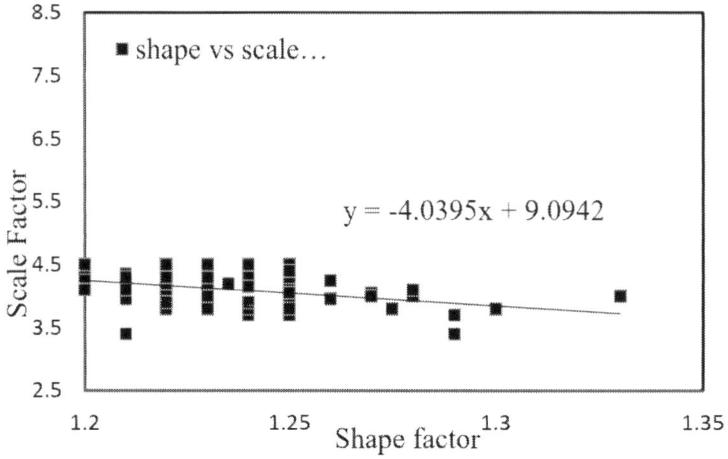

(a)

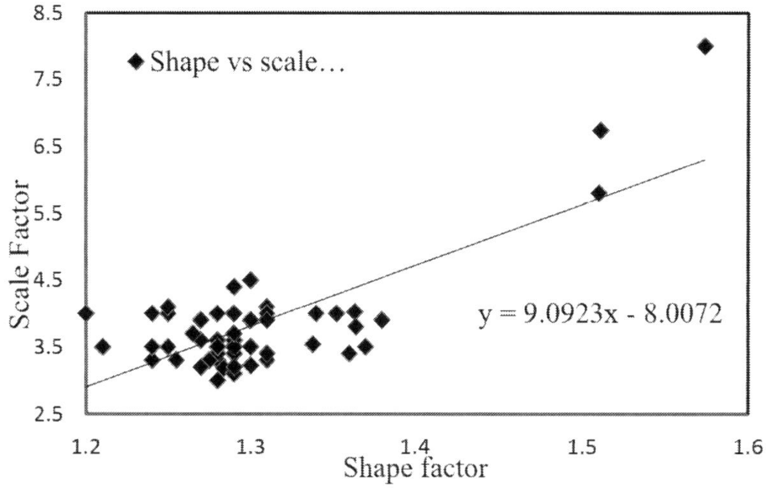

(b)

Figure 8: Correlation between Shape and Scale parameter (a) for CWTT (b) for APA

It is inferred from the Figure 7 that β for CWTT is a bit more scattered than APA. Few values of CWTT are out of bound range. The reason might be that specimens tested by CWT test device were compacted using roller compactor where control of air voids is not possible. While APA specimens were compacted using super pave gyratory compactor where volumetric control is possible. Otherwise, a linear correlation has been observed between shape parameter and scale parameter in case of both CWTT and APA.

9. Model Validation:

In order to discuss the correlation among dependent and independent variables, goodness of fit of CWT and APA rut depth model exhibited in Equation 2 and 3 is evaluated at this stage. The output of results for shape and scale factor has been shown in Table 4 and 5 respectively. As shape parameter is considered to be dependent on stability of asphalt mixtures and temperature conditions and scale parameter is considered to be dependent on shape parameter, it is important to validate the statistical significance of these variables.

Table 4: Parameter Estimate of multiple regression for Shape Parameter for CWTT and APA

Parameter	Estimate	Standard Error	t- statistics	P-Value
S	-0.312	0.03	10.4	0.015
T	0.003	0.0001	34	0.001
Constant	2.075	0.32	6.5	0.000

Table 5: Parameter Estimate of multiple regression for Scale Parameter for CWTT and APA

Parameter	Estimate	Standard Error	t- statistics	P-Value
Shape	3.98	0.08	49.75	0.000
Constant	-1.10	0.12	9.16	0.034

The Estimate column in Table 3 and 4 represents the coefficient of the respective variables. Standard error column shows the accuracy of calculations being made. A t-statistic is calculated by dividing the estimated value by its standard error which is then compared with t-critical value which is determined from selected confidence level and degree of freedom. Confidence level of 95% has been selected. The P value in last column tells us whether a variable is statistically significant or not. Since the P-value in table is less than 0.05, there is a statistically significant relationship between the variables at the 95.0% confidence level.

Now as the predicted models for shape and scale parameters have been verified, these values are used to find rut depth based on Weibull distribution by putting values of shape and scale parameter obtained from Equation 2 and 3 respectively, in Equation 1. Rut depth is the function of temperature and stability only.as stability is the indicator of both aggregates and bitumen properties, so it is convenient way of analysing different aggregates and bitumen sources. Figure 9 shows the correlation obtained between actual rut depth value and rut depth for CWTT as predicted form Equation-1

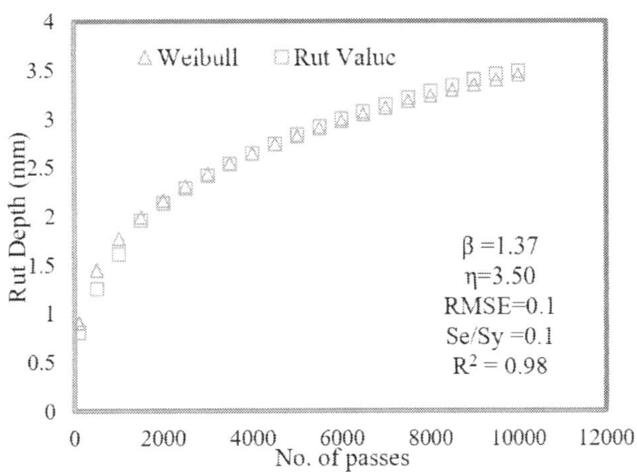

(a)

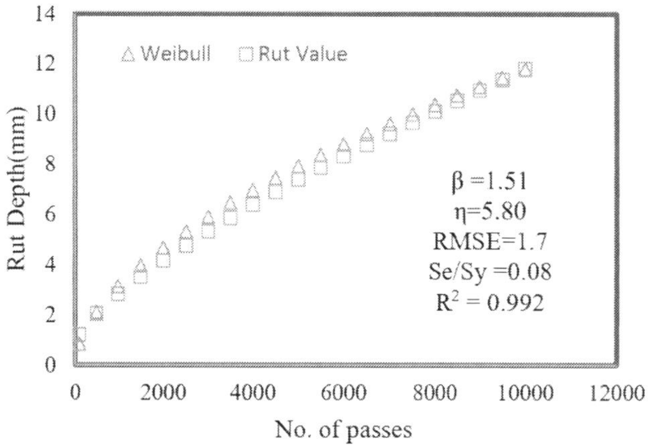

(b)

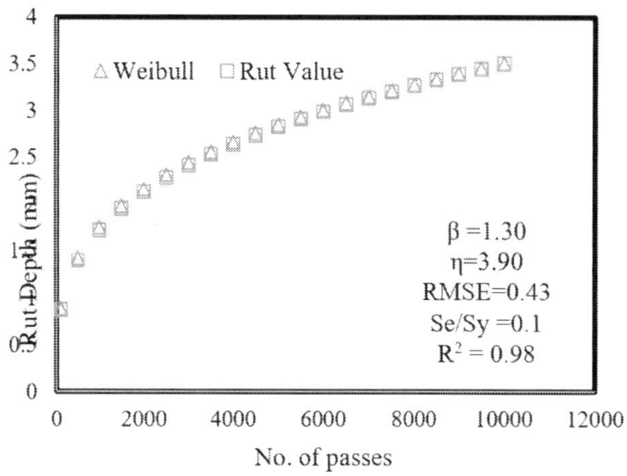

(c)

Figure 9: Actual VS Predicted Rut depth for CWTT (a) Uban Shah NRL 60/70 (b) Margalla NRL 40/50 (c) Sargodha NRL 40/50

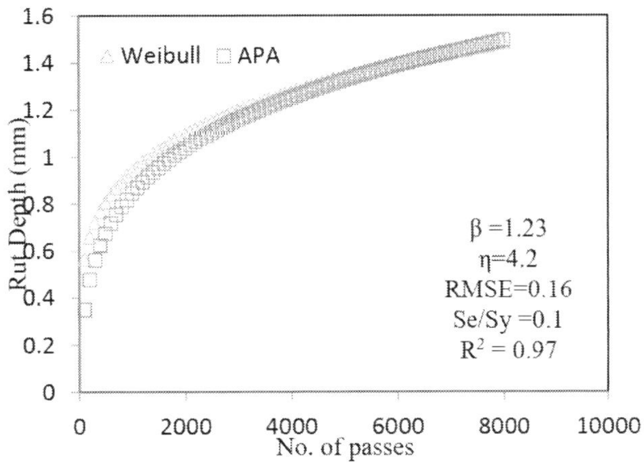
Weibull
APA
Rut Depth (mm)
No. of passes
β =1.23
η=4.2
RMSE=0.16
Se/Sy =0.1
R^2 = 0.97

(a)

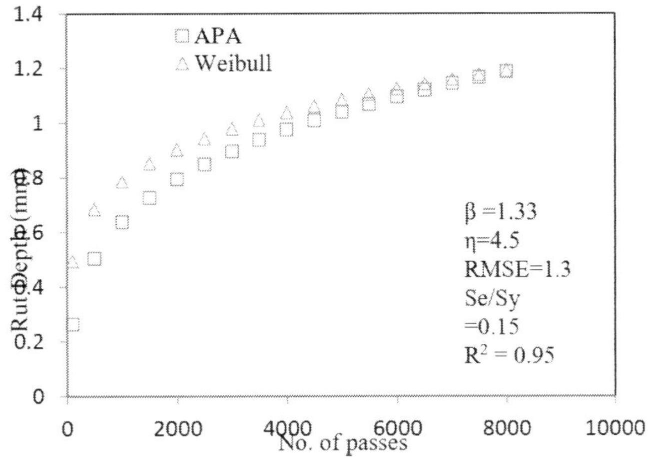
APA
Weibull
Rut Depth (mm)
No. of passes
β =1.33
η=4.5
RMSE=1.3
Se/Sy
=0.15
R^2 = 0.95

(b)

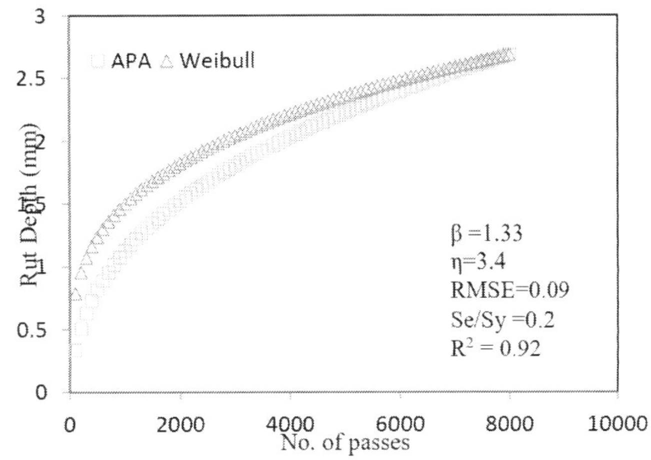
APA
Weibull
Rut Depth (mm)
No. of passes
β =1.33
η=3.4
RMSE=0.09
Se/Sy =0.2
R^2 = 0.92

(c)

IOP Conf. Series: Materials Science and Engineering **899** (2020) 012017 doi:10.1088/1757-899X/899/1/012017

Figure 10: Actual VS Predicted Rut depth for APA (a) Sargodha NRL 40/50(b)Uban Shah NRL 60/70 (c)Margalla ARL 60/70

A good correlation has been observed between actual and predicted values as translated from Figure 9 and 10. It has been observed that same Weibull distribution model predicts both APA and CWTT data with precision.

A correlation has been developed between CWTT and APA rut depth as shown in Figure 11. Quadratic equation correlates both values well with R2 value of 0.86 for 50 ℃ temperature and 0.84 for 40° C temperature.

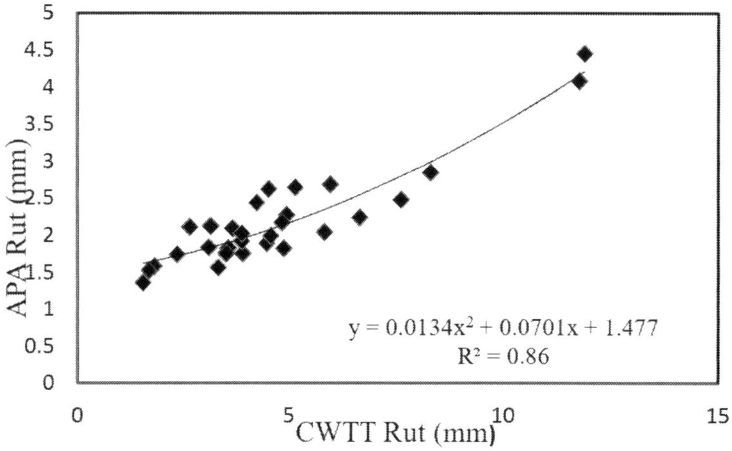

(a)

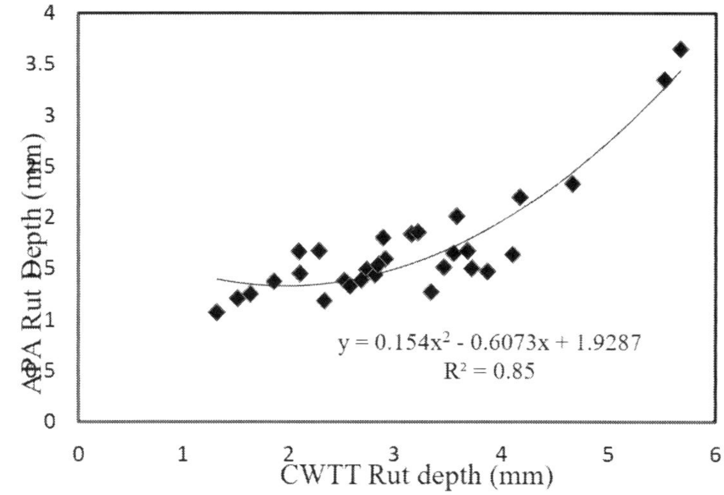

(b)

Figure 11: Correlation of CWTT and APA (a) at 50 C (b) at 40C

It can be seen from the above figures that rut values are higher at 50℃ temperature as compared to 40℃ for both CWTT and APA. It is also observed that rut values are comparatively lower for APA. This is because of the fact that loading frequency is less in case CWTT than APA, slower the frequency of load, greater is the permanent deformation of asphalt mix. The correlation between CWTT and APA recommends that instead of dealing with the costly and complicated rut simulation technique of APA, CWTT can be used as an alternative tool.

10. Conclusion:

In this research study, an effort has been made to develop a probabilistic model to ascertain the rutting resistance of asphalt concrete mixtures prepared from using varying gradations of aggregates and three distinct types of binder and aggregate samples. Results of CWT and APA has been modelled using 2 parameter Weibull distribution technique. Following conclusions are drawn based on the analysis and modelling of test results.

- It has been observed that rut depth variability is quite prominent in the case of Uban shah aggregates and 60/70 penetration grade bitumen because of the presence of flaky and elongated aggregates. While bitumen with penetration grade of 40/50 are more resistant towards rut phenomenon. None of the mixtures reached their failure state prior to the test conditions in both CWTT and APA.
- Weibull distribution technique can be used for better prediction of laboratory rutting. Asphalt mixture stability, temperature conditions and Number of loading cycles are dependent variables for modelling Weibull Rutting prediction. All samples have shown rut value up-to secondary phase. This is also one of the reasons that Weibull distribution modelled the rutting behaviour precisely.
- Model Validation shows that temperature condition plays a very significant role in the rutting of asphalt mixtures irrespective of the type of material being used. Smaller p value in case of temperature and stability factor shows that these variables have strong significance in affecting rutting potential of asphalt concrete mixtures.
- It is inferred that CWTT and APA rut data correlates well with R^2 value of 0.84 for 50 °C and 0.84 for 40°C.

11. References:

[1] T. Peng, X. L. Wang, and S. F. Chen, "Pavement Performance Prediction Model Based on Weibull Distribution," in *Applied mechanics and materials*, 2013, vol. 378: Trans Tech Publ, pp. 61-64.

[2] G. Dennert, M. Zwahlen, M. Brinkman, M. Vinceti, M. P. Zeegers, and M. Horneber, "Selenium for preventing cancer," *Sao Paulo Medical Journal,* vol. 130, no. 1, 2012.

[3] S. Hussan, M. A. Kamal, I. Hafeez, and N. Ahmad, "Evaluation and modelling of permanent deformation behaviour of asphalt mixtures using dynamic creep test in uniaxial mode," *International Journal of Pavement Engineering,* vol. 20, no. 9, pp. 1026-1043, 2019.

[4] S. Hussan, M. A. Kamal, I. Hafeez, N. Ahmad, S. Khanzada, and S. Ahmed, "Modelling asphalt pavement analyzer rut depth using different statistical techniques," *Road Materials and Pavement Design,* vol. 21, no. 1, pp. 117-142, 2020.

[5] S. Hussan, M. A. Kamal, and I. Hafeez, "Modeling and correlating rut depth observed in different asphalt mixture performance tests," *Mehran University Research Journal of Engineering and Technology,* vol. 37, no. 3, pp. 553-570, 2018.

[6] P. Singh and A. K. Swamy, "Probabilistic approach to characterise laboratory rutting behaviour of asphalt concrete mixtures," *International Journal of Pavement Engineering,* vol. 21, no. 3, pp. 384-396, 2020.

[7] I. S. J. Al-Haydari, "A Neuro-Fuzzy and Neural Network Approach for Rutting Potential Prediction of Asphalt Mixture Based on Creep Test," *Al-Nahrain Journal for Engineering Sciences,* vol. 21, no. 2, pp. 275-284, 2018.

[8] Q. Xu and L. N. Mohammad, "Modeling asphalt pavement rutting under accelerated testing," *Road materials and pavement design,* vol. 9, no. 4, pp. 665-687, 2008.

[9] C. Monismith, J. Deacon, and J. Harvey, "WesTrack: Performance models for permanent deformation and fatigue," *Pavement Research Center, University of California, Berkeley, June,* 2000.
[10] G. Lian-Sheng, D. Han-Cheng, and C. Jia-Qi, "Research on predicting the rutting of asphalt pavement based on a simplified burgers creep model," *Mathematical Problems in Engineering,* vol. 2017, 2017.
[11] A. Collop, D. Cebon, and M. Hardy, "Viscoelastic approach to rutting in flexible pavements," *Journal of Transportation Engineering,* vol. 121, no. 1, pp. 82-93, 1995.
[12] Z. Suo and W. G. Wong, "Nonlinear properties analysis on rutting behaviour of bituminous materials with different air void contents," *Construction and Building materials,* vol. 23, no. 12, pp. 3492-3498, 2009.
[13] J. Anochie-Boateng and J. Maina, "Permanent deformation testing for a new South African mechanistic pavement design method," *Construction and Building Materials,* vol. 26, no. 1, pp. 541-546, 2012.
[14] F. Zhou, T. Scullion, and L. Sun, "Verification and modeling of three-stage permanent deformation behavior of asphalt mixes," *Journal of Transportation Engineering,* vol. 130, no. 4, pp. 486-494, 2004.
[15] A. H. Gandomi, A. H. Alavi, M. R. Mirzahosseini, and F. M. Nejad, "Nonlinear genetic-based models for prediction of flow number of asphalt mixtures," *Journal of Materials in Civil Engineering,* vol. 23, no. 3, pp. 248-263, 2011.
[16] C.-D. Lai, D. Murthy, and M. Xie, "Weibull distributions and their applications," 2006.
[17] N. Bala and M. Napiah, "Fatigue life and rutting performance modelling of nanosilica/polymer composite modified asphalt mixtures using Weibull distribution," *International Journal of Pavement Engineering,* vol. 21, no. 4, pp. 497-506, 2020.
[18] A. A. Rahman, M. M. Mendez Larrain, and R. A. Tarefder, "Development of a nonlinear rutting model for asphalt concrete based on Weibull parameters," *International Journal of Pavement Engineering,* vol. 20, no. 9, pp. 1055-1064, 2019.
[19] M. M. M. Larrain and R. A. Tarefder, "Weibull model for rutting prediction of warm-mix asphalt agents," *Transportation research record: journal of transportation research board,* no. 2575, pp. 206-212, 2016.
[20] E. Coleri, B.-W. Tsai, and C. L. Monismith, "Pavement rutting performance prediction by integrated Weibull approach," *Transportation research record,* vol. 2087, no. 1, pp. 120-130, 2008.
[21] P. S. Kandhal and L. A. Cooley, *Accelerated laboratory rutting tests: Evaluation of the asphalt pavement analyzer*. Transportation Research Board, 2003.
[22] B. Choubane, G. C. Page, and J. A. Musselman, "Suitability of asphalt pavement analyzer for predicting pavement rutting," *Transportation Research Record,* vol. 1723, no. 1, pp. 107-115, 2000.
[23] F. Lekarp, U. Isacsson, and A. Dawson, "State of the art. II: Permanent strain response of unbound aggregates," *Journal of transportation engineering,* vol. 126, no. 1, pp. 76-83, 2000.

Indigenous Design and Development of Split Hopkinson Pressure Bar (SHPB) test setup for characterization of materials at high strain rates

M Bilal Nutkani[1,a] ,Muhammad Abid[2,b], Riffat Asim Pasha[3,a] , Uzair Ahmed Dar,[4,b]

[a]*Department of Mechanical Engineering, UET, Taxila*
[b]*Department of Mechanical Engineering, COMSAT S University Islamabad, Wah Campus*

Email address: [a)] bilalnutkani@ciitwah.edu.pk , [b)] drabid@ciitwah.edu.pk [c)] asim.pasha@uettaxila.edu.pk

Abstract. This paper presents design and development of split Hopkinson pressure bar (SHPB) indigenously for testing of different materials at high deformation rate. SHPB is capable to measure the deformation rate 100 $^{s-1}$ to 2,000 $^{s-1}$, by maintaining the uniform and uni-axial stress in the samples. SHPB apparatus consist of the mechanical components, support structure, bars supports and the incident and transmission bars, striker bar, gas gun, piston and rod assembly. The momentum trap system is used to absorb the energy in the form of friction and heat. All components are designed, manufactured, calibrated and assembled to test the samples. Results obtained at high deformation rates are used to characterize samples of different materials by converting in the form of strain, strain rate and stress amplitude in samples.

Keywords—*SHPB, High deformation rates, Stress amplitude, Stress wave speed, Impact strength, Uni-axial stress.*

1. INTRODUCTION

SHPB technique is widely used to measure high deformation and the material characteristics under the impact loading [1], [2]. SHPB in general consists of torsional, tensile and compression split Hopkinson's bar arrangements to test samples .Stress pulse in sample transferred through different mechanism i.e. with impactor which further transfer stress wave through transmission bar producing deformation in sample [3]. Result and conclusion obtained these strain pulses characterize of samples of different material are determined by converting them in strain, strain rate and stress amplitude of samples [3]. At the free surface of bars the only longitudinal wave transfers, while both the normal and shear stresses do not appear in the bar [4]. Pochhmammer [5] and Chree [6] described the longitudinal wave equations and their solution in the infinite elastic bars. Zhao and Gary [7] extended these solutions for viscoelastic materials. Skalak [8] and

Folk et. al [9] studied the response of semi-infinite elastic bar under the pressure wave. Waves travels in bars with velocity Co as described in the equation of stress wave. When the bar length to diameter ratio is greater than 10, the stress waves considered in one direction [4]. Davies R. M determined that in one dimensional wave stress in bar at any point of cross section, the distance of longitude is same and the radial force is zero [10]. Tensile test at SHPB is relatively tricky as compared to the compression tests due tensile pulse passing through sample. By designing a specimen of tensile test same test setup of SHPB can use for the tensile. SHPB produces a pressure wave which passes through incident and transmission bars. By introducing minor changes in the compression test setup, same test apparatus is used to measure the tensile properties of specimen. Lindholm and yeakley [11] introduced a hat form test specimen to use it for tension test on same test rig of SHPB instead of compression test between incident and transmission bars. Output bar in this case is of larger diameter than the input bar with hollow inside diameter to hold the specimen. Second method of tension test introduced by Nicholas [12] in which concept used i.e. compression wave reflection as tensile wave at free end [6]. First compression wave passed without affecting the specimen however when it reflects back it load the bar as tensile Nicholas [12] . This tensile wave used to characterize the sample for tensile test. Researchers have worked on the tension of input bar which leads to direct tensioning of sample. Very first method introduced by Harding et.al [13] in 1960 for direct tensile tests on SHPB setup. Harding and welsh [14] first studied the properties under tensile mechanism. However Nicholas [12] found the drawbacks in Harding method, that the rise time of striker is less which leads to imperfection of result of SHPB. And second drawback is the tensile sample is placed inside the tube. In later work of Harding and welsh [14] they eliminate the problem by strain gauging the transmission bar to measure wave. Most common method used is when hollow striker tube slides around input bar. A hollow striker hits the flange of input bar which directly produces the tensile pulse in the sample. Stress pulse is controlled by the pressure of gas which controls the velocity of striker bar. Further it helps to ensure the visibility of test specimen. This is the recent work presented on SHPB by Thakur [15], Ogawa [16], in 1984. In this arrangement there is a practical problem that the striker bar is hollow and slides over the input bar [17] so strain gauging is problem on input bar. Van Slycken et.al [8] introduced the striker bar in eccentric bar with respect to the input bar. Length of striker bar is observed as limited, as on striking it creates elastic sagging or deformation on the input bar as Gerlach et.al [18] stated in his work. A hollow U-shaped striker of maximum length is introduced by Gerlach which impact on flange of the input bar, sagging in the input bar due to long striker bar is controlled by supports under the bar in his case. Longer striker bar helps to generate greater pulse time which can use to characterize the ductile material under long pulse time test on low strain rate [18].

2. THEORITICAL

When material is rapidly loaded or distorted at some point, the energy imparted into the material by the source of distortion can be transmitted in the form of waves. A wave is a disturbance that propagates through or on the surface of a medium. These waves/stress pulse are function of both time and position, and are not transmitted instantaneously i.e. one portion of material is stressed while the other portion has not experienced this stress at given instant When force is applied on the end of the bar it travels particle to particle in the solid bar. Stress wave passes at the very high speed under force applied on the end of the bar [19].

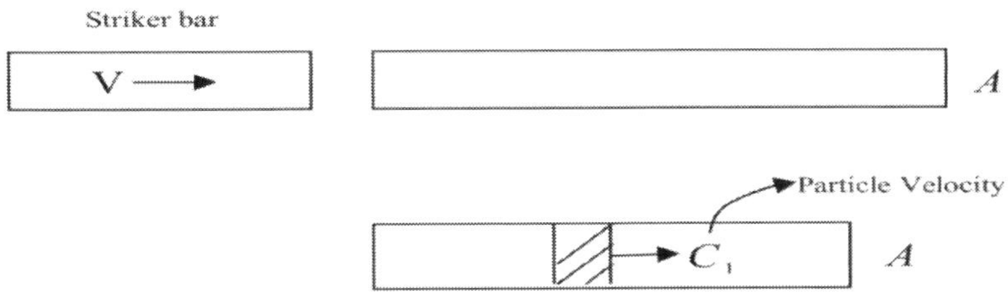

Figure 1. Stress wave propagation in bar

Permanent deformation produced in material if these waves are in plastic range and overcome the yield strength of material. Stress wave velocity is dependent on the mass, density and young's modulus of elasticity of the material (1).

$$Co = \sqrt{C}/\rho \quad (1)$$

This stress wave speed is used to measure the stress, strain and strain rate of sample as shown in (8),(9), (10).

3. SHPB EXPERIMENTAL SETUP

3.1. Design and construction of components

Design and Construction of SHB was major work to test the shape memory alloys, metallic lattice and hybrid structure At high strain rate. The main parts of the SHPB apparatus consist of the striker, incident and transmission bars, gas gun, piston rod assembly and the momentum trap system. The striker bar is used to strike the incident bar. The pulse passes through the specimen and transmitted bar at high deformation rate to characterize the material while part of wave is reflected back as a e wave. SHPB schematic is shown in Figure 2.

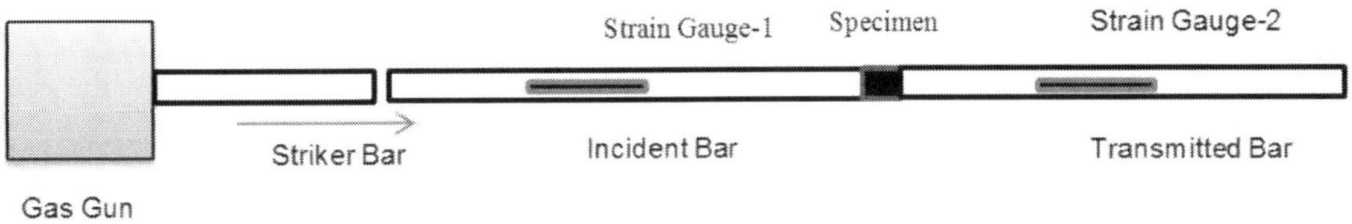

Figure 2. Schematic diagram of SHPB

To avoid the plastic deformation in bar, the striker bar must have a good resistance against the impact. AISI 4140 material due to its good fatigue strength and the abrasion impact strength, [20] [20] was selected and was machined from the round bar. Chemical composition and physical properties of AISI 4140 is provided [21][21] in the Table 1 and Table 2 respectively.

IOP Conf. Series: Materials Science and Engineering **899** (2020) 012018 doi:10.1088/1757-899X/899/1/012018

TABLE 1. Chemical composition of AISI 4140 [21].

Grade	C	P (max)	Si	S (max)	Cr	Mg	Mo
AISI 4140	0.38-0.43	0.035	0.15-0.35	0.040	0.80-1.10	0.75-1	0.15-0.25

TABLE 2. Physical properties of AISI 4140 [21][21].

Density (Kg/m^3)	Yield strength 0.2% (MPa)	Ultimate Tensile Strength (MPa)	Young's Modulus (GPa)	Elongation at Break (%)
7850	415-1735	655-1965	205	11-25.7

A same velocity is required for stress wave passing through the striker, incident and transmission bars, therefore same material is used for incident and transmission bars. Incident and transmission bar was machined identically to get the same set of bars for smooth stress wave propagation. Same lengths of both the incident and transmission bars 1830 mm were used. Length to diameter ratio of incident and transmission bars was achieved 96 to ensure the one dimensional stress wave [20]. Velocity of stress wave is calculated by equation using equation (1), AISI 4140 material used with values of $E = 205GPa$ and Density $\rho = 7850kg/m^3$, the value of $C0$ is 5110 m/s. Time of wave in bars and specimen is important.

$$T_{SB} = 2L/C_0. \qquad (2)$$

where, T_{SB} is Time for stress wave in Striker Bar, L is length of bar. With $L = 0.455$ and $C0 = 5110$, time $T_{SB} = 165\mu s$.

$$T_{IB} = L/C_0. \qquad (3)$$

where, T_{IB} is Time for stress wave of incident/transmission Bar. With $L = 1.830$m and $C_0 = 5110$, time $T_{IB} = 358\mu s$. Velocity of particles in input bar depends upon the velocity of striker bar and acoustic impedance of both striker z_S, and z_B and input bar [19] and calculated using following equation (4).

$$V_b = V_S\, Z_S\, /\, Z_B + Z_S. \qquad (4)$$

The impedance of material calculated using the following equation (5)

$$Z = \rho_0 C_0 V_b \qquad (5)$$

where A is the area of cross-section, ρ is the density of material and C_0 is the wave speed. The amplitude of stress wave depends upon velocity of striker in input bar and specimen obtained using equation (6).

$$\sigma_b = \rho A C_0 V_b \qquad (6)$$

The gas gun is used in laboratory for experimental purpose and to accelerate a projectile for Hopkinson pressure bar apparatus. We have used gas pressure because it is safer and we can attain good velocity of the projectile. Thickness of pressure chamber is calculated using equation (7) from ASME codes [22] as follows:

$$\text{tc} = \frac{PR}{SJ - 0.6P} \tag{7}$$

where *S*, *J*, *P* and *R* are allowable stress, joint efficiency, pressure in chamber and inside radius of chamber respectively. This gives the wall thickness of pressure chamber of 4mm but we use of 12mm wall thickness by using safety factor and ease of machining. To calculate thickness of wall following equation is used from ASME codes [22].Thickness of end caps are 20.8mm and 25mm standard size plates was used.

3.2. Assembly of components

Gas gun is placed on I-beam in a position that the incident and transmission bars will remain in contact and concentric. The gas gun consist major components of pressure chamber, a barrel, piston and rod arrangement, valve assembly. The pressure is developed by the gas cylinder and is stored in the pressure chamber. Two rings of Teflon of diameter 38mm was press fitted on the bar, so as to have slip fit inside the barrel. Material for the rings was selected due to low friction and light weight which keep the striker bar coaxial center to strike on the incident bar. Momentum trap system is fixed at end I-beam with screws to stop the transmission bar acting a fixed end and absorb energy in form of heat. Developed components and assembly of SHPB are shown figure 3.

3.3. Sample Design Requirement For SHPB

Literature shows that design of specimen for SHPB varies upon design of Hopkinson bar apparatus used. Two basic requirements for sample are uniaxial stress and small gauge length to reduce inertial effect [23]. Sample involves the stress wave to deform the sample that's why many limitations occur. One major work is the design and selection of geometry and clamping of sample with bars. High deformation rate in sample can obtain for circular samples in the bars. Aspect ratio of sample L/w recommended greater than 10 by different authors [24] [25] [26]. Quasi static equilibrium is basic requirement of the Split Hopkinson Bar dynamically loaded sample to achieve the proper stress and strain in the sample. Dimension of sample is important in gauge section to collect the stress, strain result distribution. Various shear test specimen also prepared as complex geometries for SHPB. Bao and Weirzbicki [27] developed butterfly shaped samples for shear tests of Aluminum alloy.

4. EXPERIMENTAL SETUP VALIDATION

One dimensional stress wave is requirement for validation of this machine to perform the tests under dynamic condition to pass stress wave equilibrium. Distance-time x-t diagram is visualization of the transmitted and reflected waves on the graph. Distance-time x-t diagram (figure 5) is developed using time of stress waves and length calculated for striker and incident bars. On vertical axis time of stress waves and horizontal axis length of bars were plotted. This x-t diagram shows incident and reflected wave passes through sample with difference of time to avoid the overlapping of compressive and tensile waves with time shift. Incident and reflected strain pulses are on upper trace while the transmitted strain pulses shown on the lower trace in x-t diagram [19]. Solid lines show the compression strain pulses while dotted line shows strain pulses which become tensile wave. By using these three times based waves measured by strain gauges we can calculate the strain rate, strain and stress for given samples. Equations (8),(9), (10) used to find the strain, strain rate and stress of sample.

$$\varepsilon(t) = \frac{-2c}{ls}\,\varepsilon_R(t) \quad (8)$$

$$\varepsilon(t) = \frac{-2c}{ls}\,\varepsilon_R(t) \quad (9)$$

$$\sigma(t) = \frac{AE_B}{A_S}\,\varepsilon_T(t) \quad (10)$$

where L_S, A, A_S, ε and σ are length of sample, area of bar, area of sample , strain pulse and stress in sample respectively.

5. CONCLUSION

The work presented indigenous design and development of Split Hopkinson Pressure Bar for testing of specimens under high deformation rate. One dimensional equilibrium of stress wave is achieved throughout the bars and sample sandwiched between incident and transmission bars. The gas gun's chamber make from a solid seamless cylinder for high pressure requirement for the setup. Incident transmission and striker, all three bars are machined of material selected for required setup. Teflon rings included on striker bar to get the smooth projectile motion to impact on incident bar. All components endogenously designed manufactured.

6. Acknowledgments

I would like to thank Prof. Dr. M. Abid for his continued support and guidance in completing this project. He is a great professor to have as an advisor and mentor throughout my education at the COMSATS University Islamabad Wah Cantt. I would like to thank the Dr. Irfan Muftay (UET Pwr) and Dr. Uzair Ahmed for their suggestions and help. I specially thanks my all colleague's including Dr. Ali Usman, Dr. Shoaib for feedback throughout this project. And Lab assistants which help me in lab during development of parts. I acknowledge PSF for funding source of this project.

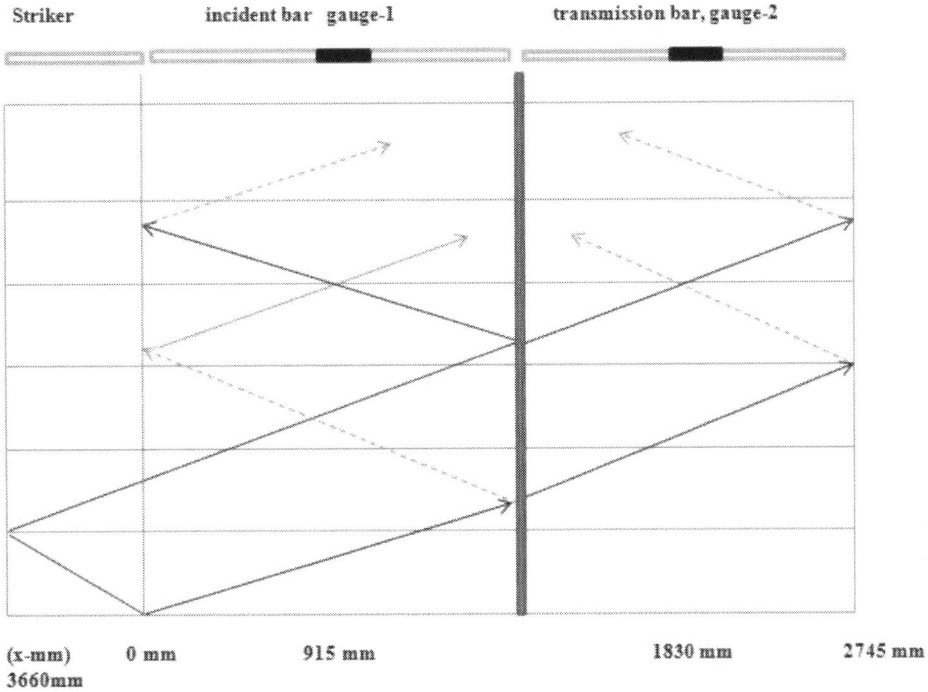

Figure 4. Lagrangian x-t diagram for Validation of SHPB bars lengths and time Calculations.

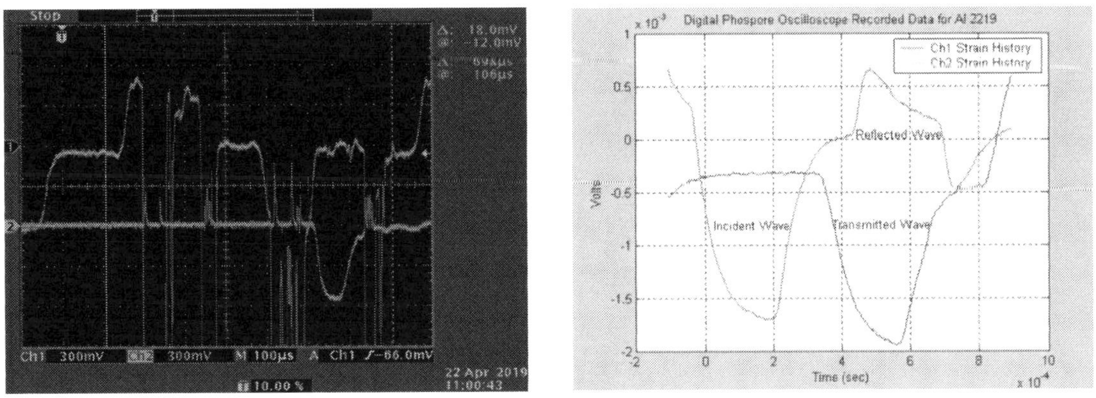

Figure 5. incident, transmitted and reflected strain pulses.

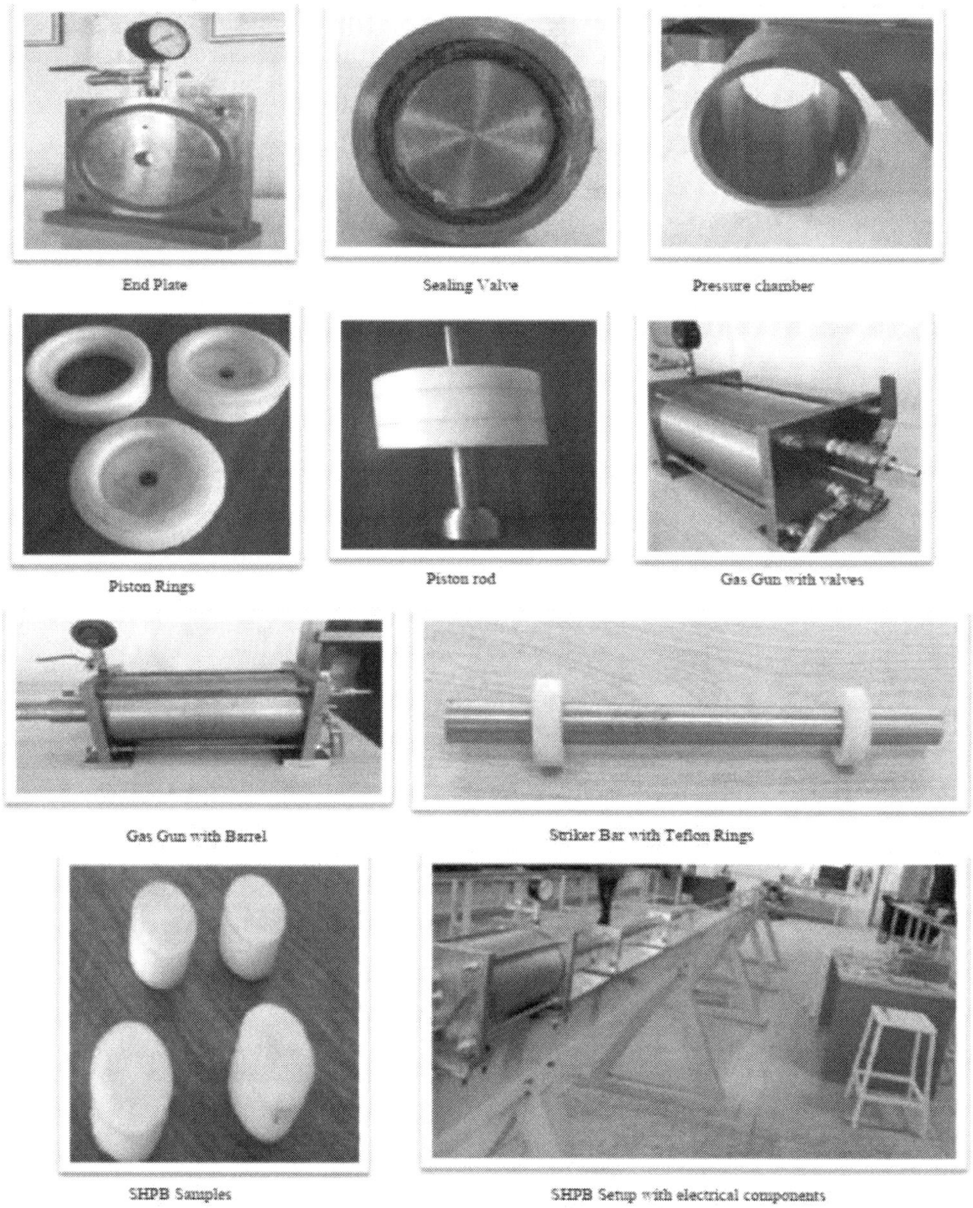

Figure 3. Indigenous Development of SHPB critical components and samples

References

[1] G. T. GRAY III, "Classic split hopkinson pressure bar testing," *ASM handbook*, vol. **8**, pp. 462–476, 2000.

[2] J. E. Field, t. M. Walley, W. Proud, H. Goldrein, and C. Siviour, "Review of experimental techniques for high rate deformation and shock studies," *International journal of impact engineering*, vol. **30**, no. 7, pp. 725–775, 2004.

[3] W. W. Chen and B. Song, *Split Hopkinson (Kolsky) bar: design, testing and applications*. Springer Science & Business Media, 2010.

[4] H. Kolsky, "Stress waves in solids," *Journal of sound and Vibration*, vol. 1, no. 1, pp. 88–1 10,1964.

[5] L. Pochhammer, "Beitrag zur theorie der biegung des kreis cylinders." *Journal fur die reine und angewandte Mathematik* ¨, vol. **81**, pp. 33–61, 1876.

[6] C. Chree, "The equations of an isotropic elastic solid in polar and cylindrical co-ordinates their solution and application," *Transactions of the Cambridge Philosophical Society*, vol. 14, p. 250, 1889.

[7] H. Zhao and G. Gary, "A three dimensional analytical solution of the longitudinal wavepropagation in an infifinite linear viscoelastic cylindrical bar. application to experimental techniques," *Journal of the Mechanics and Physics of Solids*, vol. **43**, no. 8, pp. 1335– 1348, 1995.

[8] R. Skalak, "Longitudinal impact of a semi-infifinite circular elastic bar," *J. Appl. Mech., Trans. ASME*, vol. 24, pp. 59–64, 1957.

[9] R. Folk, G. Fox, C. Shook, and C. Curtis, "Elastic strain produced by sudden application ofpressure to one end of a cylindrical bar. i. theory," *The Journal of the Acoustical Society of America*, vol. **30**, no. 6, pp. 552–558, 1958.

[10] R. Davies, "Stress waves in solids," *British journal of applied physics*, vol. **7**, no. 6, p. 203, 1956.

[11] U. Lindholm and L. Yeakley, "High strain-rate testing: tension and compression," *Experimental Mechanics*, vol. **8**, no. 1, pp. 1–9, 1968.

[12] T. Nicholas, "Tensile testing of materials at high rates of strain," *Experimental mechanics*, vol. **21**, no. 5, pp. 177–185, 1981.

[13] J. Harding, E. Wood, and J. Campbell, "Tensile testing of materials at impact rates of strain," *Journal of Mechanical Engineering Science*, vol. **2**, no. 2, pp. 88–96, 1960.

[14] J. Harding and L. M. Welsh, "A tensile testing technique for fifibre-reinforced composites at impact rates of strain," *Journal of Materials Science*, vol. 18, no. 6, pp. 1810–1826, 1983.

[15] A. Thakur, S. Nemat-Nasser, and K. Vecchio, "Dynamic bauschinger effect," *Acta materialia*, vol. **44**, no. 7, pp. 2797–2807, 1996.

[16] K. Ogawa, "Impact-tension compression test by using a split-hopkinson bar," *Experimental Mechanics*, vol. 24, no. 2, pp. 81–86, 1984.

[17] J. Van Slycken, P. Verleysen, J. Degrieck, J. Bouquerel, and B. De Cooman, "Dynamic response of aluminium containing trip steel and its constituent phases," *Materials Science and Engineering: A*, vol. **460**, pp. 516–524, 2007.

[18] R. Gerlach, C. Kettenbeil, and N. Petrinic, "A new split hopkinson tensile bar design," *International Journal of Impact Engineering*, vol. **50**, pp. 63–67, 2012.

[19] M. A. Meyers, *Dynamic behavior of materials*. John wiley & sons, 1994.

[20] K. Young, "Development of a tensile split hopkinson pressure bar testing facility," 2015.

[21] ASTM, "Standard specifification for general requirements for steel bars, carbon and alloy, hot-wrought," 2012.

[22] J. P. Ellenberger, R. Chuse, and B. E. Carson Sr, *Pressure vessels*. McGraw-Hill,, 2004.

[23] R. Smerd, S. Winkler, C. Salisbury, M. Worswick, D. Lloyd, and M. Finn, "High strain rate tensile testing of automotive aluminum alloy sheet," *International Journal of Impact Engineering*, vol. **32**, no. 1-4, pp. 541–560, 2005.

[24] J. R. Klepaczko, H. V. Nguyen, and W. K. Nowacki, "Quasi-static and dynamic shearing of sheet metals," *European Journal of Mechanics-A/Solids*, vol. **18**, no. 2, pp. 271–289, 1999.

[25] B. Bacroix, P. Genevois, and C. Teodosiu, "Plastic anisotropy in low carbon steels subjected to simple shear with strain path changes," *European journal of mechanics. A. Solids*, vol. **13**, no. 5, pp. 661–675, 1994.

[26] A. Ziolkowski, "Simple shear test in identifification of constitutive behaviour of materials submitted to large deformations-hyperelastic materials case," *Engineering Transactions*, vol. **54**, no. 4, pp. 251–269, 2006.

[27] Y. Bao and T. Wierzbicki, "On fracture locus in the equivalent strain and stress triaxiality space," *International Journal of Mechanical Sciences*, vol. **46**, no. 1, pp. 81–98, 2004.

AUTHOR INDEX

Abbasi, Hassaan ... 19
Abid, Muhammad ... 165
Afaq, Kamran S. ... 119
Ahmad, A ... 127
Ahmad, Faraz ... 40, 60
Ahmad, F ... 89
Ahmad, Hussain ... 40
Ahmad, Iftikhar ... 80
Ali, Nouman ... 100
Ali, S ... 89
Ashfaq, Fatima ... 146
Aslam, Naila Aiman ... 100
Ayaz, Adnan ... 40, 60
Chohan, Hasan Qayyum ... 80
Dar, Uzair Ahmed ... 165
Ejaz, Naeem ... 134
Farooq, Fakeha ... 35
Farooq, M N ... 55
Fatima, Saman ... 100
Gul, Rizwan M. ... 40
Hadi, S ... 55
Hafeez, Imran ... 146
Hassan, A K ... 55
Hussan, Sabahat ... 146
Iftikhar, B ... 55
Ijaz, M Z ... 89
Irgaziev, B. F. ... 1
Javed, Ali ... 6, 19
Javed, Muhammad Sohail ... 134
Kabir, Abdul ... 1
Khan, Ahmad ... 40
Khan, A ... 89
Khan, M Y ... 139
Khan, Muhammad Khizer Ali ... 6
Kiran ... 139
Malik, M M ... 55
Malik, Sheharyar ... 112, 119
Mansoor, Mohtasshim ... 6
Mazhar, Farrukh ... 6
Mir, J ... 127
Mumtaz, T ... 28
Nabi, Jameel-Un ... 1, 35
Noman, Muhammad ... 60
Nutkani, M Bilal ... 165
Pasha, Riffat Asim ... 165
Plevris, V ... 127
Qadri, Nafees Mumtaz ... 6
Rehman, A ... 28
Rehman, M M ... 55
saher, Saim ... 40
Saher, Saim ... 60
Shehzadi, Ramoona ... 35
Siddiqi, M R ... 89
Siddiqui, N A ... 139
Sohail ... 139
Tariq, Iqra ... 100
Tarmazi, S. M. hasnain ... 40
Wajid, O ... 55
Waris, M I ... 127